Performing Arctic Sovereignty

The Arctic is 5.5 million square miles and has been inhabited by humans for thousands of years, yet it is still a frontier of development. But who owns the Arctic?

This book charts the history of performances of sovereignty over the Arctic in the policy and visual representations of the US, Canada and Russia. Focusing on narratives of the effective occupation of territory found in postage stamps, it offers a novel analysis of Arctic sovereignty. Issues such as climate change, plastics pollution and resource development continue to impact the future of this space centred around the North Pole. Who is responsible for the region? This book examines how countries have absorbed Arctic territory into their national consciousness, examining the choice of, and use of, symbols and images in postage stamps. It looks at the story of how these countries have represented their Arctic frontiers and territorial peripheries.

The book argues that the performance of policy in these regions has caused relative sovereignty to become a reality. It provides an intriguing account of how these countries have, in their distinctive ways, established, legitimised and reinforced their political authority in these regions. This book will appeal to Geographers and is recommended supplementary reading for students in political history and regional studies of the North.

Corine Wood-Donnelly is a researcher at Uppsala University, Sweden, and is also affiliated with the Scott Polar Research Institute. Her interdisciplinary research in political geography and international relations specialises on the Arctic region. She has research interests in performativity, materiality and cosmologies of sovereignty, especially related to maritime spaces.

Routledge Research in Polar Regions
Series Editor: Timothy Heleniak
Nordregio International Research Centre, Sweden

The Routledge series in Polar Regions seeks to include research and policy debates about trends and events taking place in two important world regions: the Arctic and Antarctic. Previously neglected periphery regions, with climate change, resource development and shifting geopolitics, these regions are becoming increasingly crucial to happenings outside these regions. At the same time, the economies, societies and natural environments of the Arctic are undergoing rapid change. This series seeks to draw upon fieldwork, satellite observations, archival studies and other research methods which inform about crucial developments in the Polar regions. It is interdisciplinary, drawing on the work from the social sciences and humanities, bringing together cutting-edge research in the Polar regions with the policy implications.

Arctic Sustainability Research
Past, Present and Future
Andrey N. Petrov, Shauna BurnSilver, F. Stuart Chapin III, Gail Fondahl, Jessica Graybill, Kathrin Keil, Annika E. Nilsson, Rudolf Riedlsperger, and Peter Schweitzer

Resources and Sustainable Development in the Arctic
Edited by Chris Southcott, Frances Abele, Dave Natcher, and Brenda Parlee

Performing Arctic Sovereignty
Policy and Visual Narratives
Corine Wood-Donnelly

For more information about this series, please visit: www.routledge.com/Routledge-Research-in-Polar-Regions/book-series/RRPS

Performing Arctic Sovereignty

Policy and Visual Narratives

Corine Wood-Donnelly

Routledge
Taylor & Francis Group

LONDON AND NEW YORK

First published 2019
by Routledge
2 Park Square, Milton Park, Abingdon, Oxon OX14 4RN

and by Routledge
52 Vanderbilt Avenue, New York, NY 10017

Routledge is an imprint of the Taylor & Francis Group, an informa business

© 2019 Corine Wood-Donnelly

British Library Cataloguing-in-Publication Data
A catalogue record for this book is available from the British Library

Library of Congress Cataloging-in-Publication Data
A catalog record has been requested for this book

ISBN: 978-1-138-57359-8 (hbk)
ISBN: 978-0-203-70148-5 (ebk)

Typeset in Times New Roman
by Wearset Ltd, Boldon, Tyne and Wear

Contents

Figures

Acknowledgements

This book would not exist without the generous assistance, intellectual discussions and moral support from my employers, colleagues and friends. Of primary importance were the affiliations that provided the institutional infrastructure for this research. This was first as a research associate at the Scott Polar Research Institute in the Department of Geography at the University of Cambridge. Second, was the Institute for Russian and Eurasian Studies at Uppsala University, who employed me as a postdoctoral researcher. I am pleased to acknowledge their contributions.

Many people helped me over the course of this project. There are the conversations with intellectual mentors, including my PhD supervisor Gareth Dale at Brunel University. While in Cambridge, I was encouraged by fellows Michael Bravo, Gareth Rees and Alex Jeffrey. In Uppsala, I have been supported by my colleagues Vladislava Vladimirova and Matthew Kott. Finally, there are the many peers and friends around the world who have read drafts, offered criticisms and suggested improvements. I am grateful to you all.

Part I
Setting the scene

Introduction

Consider the following two events separated by the space of a century.

For 5 April 1909, a log entry reads: 'Tomorrow if ice and weather permit I shall make a long march, "boil the kettle" midway, and try to make up the five miles lost on the 3rd.'[1] These scribbled words in the expedition diary of Rear Admiral Robert E. Peary record the day before he claims to have been the first explorer to reach the North Pole. The next day he writes:

> The Pole at last!!! The prize of three centuries, my dream and ambition for 23 years. *Mine* at last. I cannot bring myself to realize it. It all seems so simple and commonplace, as Bartlett said 'just like every day'.[2]

Aside from boiling the kettle, there was an additional ceremony to perform on reaching the North Pole. Peary recalls,

> Of course there were some more or less informal ceremonies connected with our arrival at our difficult destination, but they were not of a very elaborate character. We planted five flags at the top of the world. The first one was a silk American flag....[3]

Following this ceremony, Peary placed a fragment from the flag in a bottle, along with a note reading:

> I have to-day hoisted the national ensign of the United States of America at this place, which my observations indicate to be the North Polar axis of the earth, and have formally taken possession of the entire region, and adjacent, for and in the name of the President of the United States of America.[4]

They take photographs and the expedition records measurements of position and temperature before leaving that point so long dreamed.

When Peary returned from the North Pole, his actions were equally as symbolically significant, even if the actual performance of his activities was without ceremony. Upon reaching civilisation, he reports telegramming the US President with a message saying, 'Stars and Stripes nailed to the North Pole.'[5] According to media reports, the President replied congratulating Peary on his achievement, noting the honour it brought to his countrymen. Directly addressing the message regarding the posting of the flag, the President said, 'Thanks for your interesting and generous offer. I do not know exactly what I could do with it.'[6]

Nearly 100 years later....

For 1 August 2007, an expedition account reports that a polar bear is sighted by observers on the *Rossia* icebreaker, the first visual of the species at the extreme latitude of their position: the surface of the Arctic Ocean at the North Pole.[7] With good weather conditions on their side, the main objective for the expedition is planned for the next morning. Tomorrow will be the day when the Russian hero of polar exploration, Artur Chilingarov, attempts to achieve a new first in world history, the first time anyone has 'ever gone to the "real" North Pole', the last of the firsts in the Arctic.[8] The dive to reach the seabed is successful and Chilingarov later recalls, 'How good it is down there. If in a hundred or a thousand years to where we were, someone would come down, then he would see the Russian flag.'[9]

Aside from claiming the title of the first expedition to reach the Real North Pole in vessels from the Russian Academy of Sciences, there were additional ceremonies to perform at the site. The group places a time capsule for future generations together with a plaque and a titanium flag of the Russian Federation. The expedition report notes that these objects commemorate 'the dive as a historic Russian achievement', adding 'another chapter to their nation's history of polar exploration'.[10] The team also captures video footage, records measurements of pressure and collects soil and sea life samples for analysis.

When the expedition team returned to the surface the ceremonial aspects of the event continued. The explorers emerged from the submarine waving a Russian flag, and following, a call was made from the ship back to Russia announcing, 'the MIR 1 sub has successfully concluded its mission at the North Pole. It did achieve the seabed at the point of the geographic North Pole ... and did install the Russian state flag in titanium.'[11] When he returned to Russia, while carrying a photograph of the newly planted flag at the North Pole, Chilingarov stated, 'the Arctic has always been ours and the fact that we are there first is great'.[12] The already famous explorer was congratulated by the Russian President and later awarded the title Hero of Russia.

The symbolism of these two events at the North Pole is easy to grasp because they are repeating well understood ceremonies of possession, symbolic performances historically used by states claiming territories previously unclaimed by other sovereign states.[13] Although a century apart, the ventures led by Peary and Chilingarov have the right cocktail of ingredients for the practice of making territorial claims. There are the 'firsts', a critical element of claiming new territory as one cannot assert title of a territory already in the possession of another state. Next, there is the recording of performances, chronicling the placing of the national flags or other memorabilia, along with communications identifying when and how the possession occurred. Finally, there is the communication of the success of the acquisition process back to state authorities and to the nation.

From their beginnings in the fifteenth century, the repetitions of these performances is an explanation for 'how the west was won', even if only theoretically, and often irrespective of the already existing populations. It is the way in which states asserted their moral authority over a geographic space. Emerging from the explorations and 'discoveries' of Columbus, imperial states tacitly assented through repetition of practices, thereby establishing procedural norms for gaining territory beyond the existing national territory. The result is that the processes for claiming new territory became an established set of performative rules: discovery, ceremonies of possession, communication and then finally, one last element not listed above: effective occupation.[14]

Effective occupation came to mean the implementation of the political technologies and systemic structures of the state: government administration, colonialism, resource management, taxation, and civilizing policies for indigenous communities as a list of examples. For the Arctic, the final solidifying element of effective occupation has been an elusive target given the hostile conditions of the Arctic climate. In fact, because states had been so ineffective in the execution of occupation of their Arctic territories, legal scholars came to give the Arctic region a get-out principle to allow for the lack of effective occupation, avoiding declaring the Arctic as open to another cycle of new sovereignty claims. So for the Arctic, effective occupation came to mean as much occupation as can be reasonably exercised given the extreme climes.[15]

Despite the general attitude that administration requirements were less stringent for maintaining sovereignty in the Arctic and while most states accepted that other Arctic states were meeting satisfactory standards, the ongoing lack of complete effective occupation had two effects on the behaviour of states. First, for some states it introduced an internal crisis of confidence in their unsubstantiated claims to Arctic sovereignty. In the Arctic, some of the territories have been incorporated into national maps without the usual procedures from first discovery to effective occupation. This includes claiming sovereignty without the privilege of first discovery, to the absence of state-sponsored territorial survey exercises, normally considered as part of performance cartography and the transformation of space into place.[16] Instead, some claims to Arctic sovereignty were perpetuated by mere declarations or speech acts, as in the case of the sector claims of Russia and Canada.

Second, the absence of ordinary administration structures compelled states to become creative in their interpretation and performance of effective occupation. The range of strategies include everything from the government production of maps illustrating the boundaries of sovereignty to relocation polices for indigenous peoples. This creativity continues in the mechanisms used to communicate their continued intentions towards maintaining their sovereignty in the Arctic.

Today, states with Arctic territory are communicating their sovereignty through performances of effective occupation in a more overt manner than they do with their other established territorial realms. Imagine, if the Canadian Prime Minister was making similar statements regarding Canada establishing sovereignty over Ottawa's province of Ontario, as it does over its Arctic islands! It is these performances and their communications, including diplomatic statements regarding the establishing and enforcing of Arctic sovereignty, that continue to communicate the subconscious crisis of sovereignty to observers. The communication of this crisis is a source of the prevailing public perception that the Arctic is territorial land grab. It is the variety of performances used to demonstrate Arctic sovereignty that will be the focus of this book.

Added to this crisis of confidence is the shifting nature of territorial sovereignty. Since the seventeenth century, the 'cannon shot rule' was the standard for measuring the breadth of maritime territorial sovereignty. Combined with the predominance of the principle of the 'freedom of the seas', it was inappropriate to claim sovereignty over maritime spaces beyond territorial waters of three nautical miles. This explains the US President's disinterest in Peary's offer of possession of the North Pole. Given the pole's location in open ocean spaces, the principle of the freedom of the seas disqualified it as a geographical point available for possession, even though its location was permanently frozen. Even today, ice used as land continues to evade internationally recognised legal status.

Regarding the evolution of normative rules, territorial sovereignty experienced a major shift between Peary and Chilingarov's visits to the North Pole with the additions to the law of the sea making significant advances. In the twentieth century, first through a unilateral performance and then through codification, the rules of maritime sovereignty changed, designating to states sovereign rights over the submarine continental shelf in the UN law of the sea.[17] At first, sovereign claims over submarine territory could extend up to 200 nautical miles from the shoreline. When in 1982 the law of the sea was renegotiated, the outer boundary was increased to permit claims of up to 350 nautical miles, providing that certain scientific conditions can be proven. With deadlines to submit these claims at ten years after ratification of the convention, many Arctic states have been feverishly surveying the potential extensions to their sovereign maritime territories in recent years. At this time the submitted claims and some resubmitted claims are awaiting review. It is unfortunate that these submissions coincide with a US Geological Survey report on potential oil reserves in the region, adding fuel to the speculation of an Arctic gold rush.[18]

In spite of the orderly adherence to the processes of international law by states submitting their Arctic claims to the UN Commission on the Continental Shelf, there is a juncture when this process could fail.[19] In the small Arctic Ocean, there is a great likelihood that these sovereign claims to maritime territory, extending up to 350 nautical miles, will intersect and overlap at the North Pole.[20] Given this, despite the parameters in the law for scientific measurements to determine the outer limits of sovereignty, the notion of historicity may still be a factor in determining sovereignty in the event of a dispute over sovereignty of the North Pole. Hence this performance by Russia is not an outright ceremony of possession – it is still rather a performance of effective occupation – with claims that they reached the real pole first, they left their mark, conducted experiments and communicated the event to the nation and to the world.

This heady mix of the ritual elements for claiming territorial sovereignty is the reason why states protested about the Russian performance in 2007. A century earlier, the US President dismissed the claim to the North Pole by Peary because it was over the ocean. With the exception of Peary's performance, it has now been some time since flag planting has been used to initiate territorial claims, and this is especially atypical for maritime territory; since then, the rules and norms have changed. Flag planting is not considered to be a legitimate performance for new claims, hence the Canadian Foreign Affairs Minister's dissent to Chilingarov's performance with, 'Look, this isn't the 15th century. You can't go around the world and just plant flags and say "We're claiming this territory". Our claims over the Arctic are very well established.'[21] There are, however, new rules on the performances of territorial claims involving internationally recognised legal codes. Therefore, for the present, territorial claims in the Arctic Ocean are first to be delimited through the law of the sea.

An additional aspect to consider is that as the climatic conditions become a diminishing obstacle, both due to global warming and the improvement of technological capabilities, the conceptual boundaries for 'as much as is reasonable effective occupation' will expand. This will cause the administrative functions of states claiming sovereignty in Arctic territories to improve until they align more closely with their domestic responsibilities elsewhere in the country. Together these conditions create a 'perfect storm' wherein states must continue to be creative in the ways they express sovereignty over the Arctic. Until the territorial boundary lines are drawn and accepted as legitimate claims by other states, the show and the performances of effective occupation must go on.

It is the goal of this book to demonstrate the performances of sovereignty that the Arctic states, through a variety of policies and approaches, have used to demonstrate their effective occupation over the territories of the Arctic. Through the narrative of the history of Arctic policy of the US, Canada and Russia, it will demonstrate how performative utterances presented in state issued declarations in claims to sovereignty give indicative promises to fulfil the requirements of sovereign authority in that space through the exercise of political technologies ordinarily associated with state control.[22] Leading on from this, it demonstrates how the performance of sovereignty through Arctic policy, as textualised government 'speech acts', results in performativity and the legitimisation of authority over the spaces through the performance of rule-governed behaviour and specifically over the space of the Arctic.[23]

This book argues for an understanding of sovereignty over the Arctic as a territorial authority socially constructed by states through performances, where these actions are conceived of as co-producing knowledge about this space through the exchange of symbolic performances of sovereignty. To make this argument, this book explores the evolution of Arctic policy in response to the changing norms of territorial sovereignty. To illustrate these performances and the communication of sovereign authority to a wide audience, it considers these messages as published in state documents and as communicated not only through government statements, but also as transmitted through 'the tiniest messenger of the state', the postage stamp. This approach contributes to three areas of existing scholarship.

First, this research contributes to the growing body of current scholarship on Arctic geopolitics. The importance of the Arctic, and especially the importance of Arctic sovereignty, has accelerated due to drivers of climate change and resource exploitation. However, there is a significant divide in the literature of post-Cold War studies of the Arctic region. On one hand are those who focus on a realpolitik explanation of Arctic geopolitics – the scramble for territorial stakes framed largely within narratives of conflict, military insecurity and divergent national interests.[24] On the other hand is the larger part of Arctic literature, and where this work is situated, a position that sees narratives of a resource scramble as misinformed and inadequate for explaining the complexities of the Arctic within a global context. It is not that states are not interested in Arctic resources or seeking to make certain their sovereignty in the region, but it is rather 'how the Polar Regions have been caught up in power-knowledge scrambles'.[25] This body of literature focuses on how the Arctic is situated within the context of significant global processes and provides 'critical analysis on the state of Arctic geopolitics and security in the era of globalization with complex and deeply interdependent ecological, economic, environmental, cultural, political, and societal processes'.[26] This book places itself as narrative of the social construction and evolution of Arctic sovereignty against the backdrop of the evolving international system and the rules, exigencies and concerns that order the normative changes and how these rules are applied to the Arctic.

Second, the research contributes to a new turn in studies of critical geography and demonstrations of how the applied performances of the state create the condition of sovereignty.[27] Recent scholarship in the fields of critical and political geopolitics have been 'influenced by discourse analysis and social constructivism … concerned with the political implications of cartographic representation, linguistic configurations, and rhetorical patterns'.[28] This body of research 'has looked beyond traditional preoccupation with textual analysis of policy statements to explore the forms of practices and materials that produce geopolitical knowledge', revealing how states create their own reality.[29] By exploring the performance of the norms of sovereignty in the Arctic, the argument contributes to 'emerging work studying the production of territorial knowledge',[30] and in studying the performances of sovereignty in effective occupation politics of the Arctic, it applies the idea that 'norms are the result of interest coordination'[31] established, interpreted and reinforced by state practices. Consequently, this study draws not only on an analysis of policy history, but also considers how those messages have been communicated in Arctic postage stamps.

Finally, this book contributes to the growing body of research that uses the postage stamp as a form of government discourse, the smallest messenger of the state, because they 'emerge as vehicles for identity creation and propagation, and as mechanisms for regime legitimation'.[32] As postage stamps are widely distributed, both to domestic and international audiences, 'the messages may be subconsciously and subtly conveyed, and it can be argued that the process of repetition further consolidates the process of delivering the message'.[33] Some research shows how the postage stamps and postal administration serve as a method for demonstrating domestic authority or transitions in authority and also how they are used to signal authority beyond the borders of the state.[34] In the case of this research, the postage stamps depicting the Arctic show not only the repetition of particular messages, but also that these messages coincide with the performances of sovereignty and the broader geopolitical concerns of the US, Canada and Russia with regards to Arctic spaces.

This book makes several arguments related to these areas of research. First, it argues that the Arctic is a socially constructed space and that the communication of the relationship of states to this territory has developed alongside the evolution of normative frameworks for understanding sovereignty. A second argument is that territorial authority in the Arctic is the result of performances of sovereignty, performances which communicate alignment with the understood norms of sovereignty for the period, or that they signal a shift in normative understandings. Further, as a frontier region with some indeterminate sovereignty, states communicate mixed messages through their performances which align better with past norms, causing insecurity in the Arctic region. Finally, it argues that these performances of sovereignty are evident in the postage stamps circulated by the US, Canada and Russia depicting Arctic events, environment and culture. The following sections discuss in more detail how this book contributes to these three areas of scholarship, followed by a brief summarisation of the book's structure.

On sovereignty

Sovereignty is a condition generated through the application and exercise of authority; it is the unequivocal authority of a state over a given territory and its people.[35] In an abbreviated discussion of the topic which avoids some of the pitfalls of presenting sovereignty as having fixed or stable meanings, this research presents sovereignty as both an uncontested concept for grounding the origins of state authority, but also as a concept contextually situated in time and place.[36] When depicted as a matrix, sovereignty is described as having both internal and external features. It should be understood that in some cases, that as an internal condition, sovereign authority is positioned to originate in a social contract between the state and citizens, establishing the domestic legitimacy of state authority.[37] Through this contract, the state in effect promises to ensure the security of its citizens. As an external condition, the state's legitimacy is realised through the acknowledgement of other sovereign entities. Without this recognition, an unrecognised state is not equal with other states in the international system. In both the domestic and international realms, sovereignty requires action, and as a result, it is a performance.

Sovereignty, as an ideational conceptualisation of authority, is not something that can be visually identified in a single item or legal person, such as a crown or a head of state or government, but the evidence of its existence can be seen through a multitude of symbols and objects that collectively demonstrate the existence of the authoritative body in a given space. Of course, crowns, flags and currency are identifiable as symbols of state authority within this set, but equally significant are other more mundane and everyday features of government administration: passport control at border crossings, collecting taxes, paving roads, issuing of licences for oil exploration and, of course, post boxes and postage stamps.[38] Some of these latter features, while presenting their existence with physical evidence, are only evidence of sovereign authority when they are used, or performed. After all, what use is a duty paid for postal services, represented through the presence of a postage stamp, if the infrastructure and systems needed to give value to the stamped tariff or to deliver messages to their intended destination do not exist?

In relation to external sovereignty, performances of a state's authority over a given territory can also be produced or reinforced by legal persons outside of the state; sovereignty is not strictly an outward projecting force. First, there is the need for a legitimacy of a state as a legal authority to be recognised by the international community.[39] This happens, for example, when a new state gains recognition in the international community by gaining a seat at the UN. Second, performances of sovereignty can take a bodily form. Take as an example the fact that maritime sovereignty is performed by foreign ships when they hoist a 'courtesy flag' of the state whose waters they are entering.[40] There are other examples of the performances of sovereignty by external actors in the Arctic, such as when US military vessels apply for environmental clearance with Canada to traverse the Northwest Passage or when commercial ships apply for a licence from Russia to navigate the Northern Sea Route.

Returning to the sovereignty matrix, on another axis, sovereignty can be seen as either relative or absolute, dependent upon the projected recipient or territorial target.[41] Few would deny the authority of the sovereign state to make laws for its people; they have 'relatively exclusive power to exercise domestic, transnational, and international legal authority in relations to matters that bear on its territory and population'[42] that is, until they violate the moral laws of the international community. Yet limits do exist to state sovereignty, creating a condition of relative sovereignty. Given that territorial stability in the international system is premised on the notion of the fixed borders of states, the 'significance of borders derives from the importance of territoriality as an organizing principle of political and social life'.[43] As a result, the demarcation of borders and the performance of borders becomes a significant part of defining the limits of the sovereign state. It is by consent when states enter agreements of international law as a legal person that limits absolute sovereignty, while it is the capacity of these states to extend their authority to extra-territorial spaces that creates the condition of relative sovereignty. In the Arctic, it is relative sovereignty that applies to the maritime spaces and absolute sovereignty to terra firma, creating different layers of sovereignty and a condition whereby state authority is unequivocal in some areas while in other areas it is restricted by international law.[44]

The importance of these nuanced differences between external/internal sovereignty and of absolute/relative sovereignty is significant in the way in which they affect the geographies of the Arctic and the interplay of states in the region. The Arctic is a region characterised by stratified sovereignty, initiated by the practices of imperialism and framed by evolving international law. In some of these spaces the Arctic states reproduce sovereignty by controlling the space through mechanisms of international governance and other performances of sovereignty, such as by extending environmental legislation over waters, as seen with Canada in the Northwest Passage. The effects of this stratification produce both inconsistent discourse and administrative approaches throughout the Arctic as states seek to substantiate their sovereignty through performed policy solutions in geographic areas with different normative rules organizing state behaviour, expectations and rights of states over territory and resources. The result of this evolution is that the sovereignty of the Arctic is spatially distributed over different territorial spheres, which makes for uneven distribution of government administration.

Performances of sovereignty in critical studies

The conceptual ideas for performance and performativity is an expanding body of scholarship beginning in theatre studies, but adopted by geography and political sciences. These concepts have roots in the work of Butler, who argues that gender and the identities associated with gender are not genetically determined or socially conditioned but instead are social constructions and the result of repeated performances of specific behaviours that convey to the receptor the gender identity of the actor. In the way that performances produce meaning, such as performances of gender or identity creating understandings of social realities, the performances of the state are likewise practices that produce meaning.[45] Through the repetition of certain types of behaviours and rules associated with the structure of the state, the state has the ability to perform the story of itself to both its domestic audience and to the international community, creating a semblance of legitimacy.[46]

States are socially constructed, both as political ideas and as physical entities. As a political idea, states have come into being by their ability to project power and through their ability to gain legitimacy. This construction is a process that happens through performance and through practices of the state through the repetition of rules, or norms, that the community of states recognises as legitimate behaviour. To do this, they borrow from a toolkit of recognised symbols ranging from the spectacular to the mundane – from flag waving ceremonies to the delivering of the daily post. These performances communicate compliance with these rules, that states, in order to be legitimate, must fulfil certain types of functions resulting in the administration and security of the state. These administrative performances are often what constructs the physical aspect of the state. The state comes into physical being when it has a material manifestation of the political social construction throughout a spatialized territory.

The evidence of these types of performances have been investigated by a number of scholars where 'the spatio-legal performance of that claim, adjudication, and admission/expulsion is a "deep structure" of sovereign politics'[47] serving either to reinforce or create the sovereignty of the state. For example, Brown has investigated the ways in which building walls and fences at national borders is a mechanism to 'shore up' state sovereignty. She says they are a:

> theatricalized and spectacularized performance of sovereign power at aspirational or actual national borders ... [where] If walls do not actually accomplish the interdiction fueling and legitimating them, if they perversely institutionalize the contested and degraded status of the boundaries they limn, they nevertheless state both sovereign jurisdiction and an aura of sovereign power and awe.[48]

This performance of borders has also been taken up by Weizman who shows through cartographic imaginary and the use of material architecture the way in which political performances are 'fully absorbed into the organization, transformation, erasure and subversion of space', spaces that are 'the medium that [state] actions seeks to challenge, transform or appropriate',[49] adding a vertical dimension to the sovereignty of space. Meanwhile, Busbridge claims that by delineating spaces, 'the border, as a political-geographical site, can thus be understood as a legal, social and special performance of sovereignty'.[50] These scholars demonstrate how performances of sovereignty, such as in frontier, border and vertical spaces, serve to legitimise the authority of the state in specific geospatial relations.

In other work, scholars investigate roles of the performance of the state in creating legitimacy at a more ideational level. For example, Krahmann has discussed performativity in 'a focus on the repetitive enactment of specific forms of behaviour and capabilities'[51] as a mechanism for demonstrating legitimacy through performativity in the area of public goods. Meanwhile, in the process of creating a new state in the aftermath of a civil war, Jeffrey argues that, 'States are improvised. Their legitimacy and ability to lay claim to rule rely on a capacity to perform their power'; he further posits that these 'performances are structured by available resources'.[52] Each of these are instances of states performing their sovereignty, such as in a study of writing the state through intervention practices, Weber says that 'Sovereignty [is] a code. It is a bundle of practices which, when performed, grant specific rights and responsibilities.... What become important are the signs of sovereignty – the ability to access the code of sovereignty.'[53] These works together demonstrate how states use a variety of political technologies to convey power, capacity and legitimacy to rule in socio-political contexts.

Adding to this study on the geography of political performativity in the context of Arctic sovereignty is 'exploring the various ways in which political spaces are both materially and discursively performed'[54] through a consideration of the use of the written and spoken word as tools of performance. In this instance, a performative is where 'the utterance constitutes the performance of the act named by the performative expression in the sentence'.[55] This occurs not only in spoken form – such as declarations of war or diplomatic address, but also in forms such as cartography, street signs, postage stamps and policy documents. By considering illocutionary and legislative publications of the state as speech acts where 'the issuing of the utterance is the performing of an action',[56] in this line of thought, policy documents and messages from government representatives are not just saying, but are also doing, or performing.

Speech acts are not just a process of mundane and banal actions of daily government administration, they are also a promise to change the world by moving from one condition to another. If a state accepts another state's assertions and statements that a certain condition exists, it secondarily assumes the responsibility to also change the world by mirroring these performances. In this way, performative rules and norms for behaviour are developed within the international system when 'agents construe them as promises, accept them as such, and proceed to rely on them'.[57] While states cannot claim explicit and absolute sovereignty over the entire Arctic, given the various rules for delimitation of ocean spaces, they can reproduce the idea of sovereignty through symbolic territorial performances and speech acts which communicate authority over this space thereby creating at least relative sovereignty over both horizontal and vertical territories. In the consideration of sovereignty in the Arctic, the utterances of states should

be considered as having performative capacity, not only due to their illocutionary elements, but also due to the role in which norms and the repetition of norms create stable meanings for the geopolitics of the Arctic.

Qualitative and quantitative research methodology with visual representations

This research builds on a growing body of literature with contributions
from scholars from fields across the social sciences considering the role of
postage stamps as messengers of the state. Using postage stamps as a
medium for analysing political discourse of a specific state or thematic
area is gaining in popularity because their visual aspects capture states'
positions in a way that the textual or spoken elements of policy do not,
giving additional substance to policy discourse analysis. It has been noted
that 'when states emphasize "the visual", which includes maps, postage
stamps, current and official Websites, they inform and educate their own
populations and those beyond about where they are, who they are and what
they are about'.[58] As a result, stamps reproduce and reinforce the narrative
of the state in a visual form by using symbols and images of their identity,
authority and governmental administration over a particular people and
territory.

With a direct connection between the state and the bureaucracy of
postal administration, stamps are indicative of sovereignty and effective
administration providing evidence of the production and cancellation of
postage stamps; the delivery of letters indicates the presence of government authority in specific geographic areas where these artefacts have
legitimate value at a given place and time.[59] Given that all states produce
postage stamps, they are consistent and pervasive empirical documents
available across all global political systems and they are often easier to
access than some other government materials, written in foreign languages
and frequently outside of the public domain. Useful as propaganda tools,
stamps serve as messengers 'through the dissemination of images that
suggest a familiarity of geographic spaces, repetitive use of indigenous
iconography, and pictorial representations of effective state administration
over the region',[60] in much a similar vein as used in cartographic reproduc-
tions of state territory. Postage stamps are, in effect, marketing mechanisms for the state to showcase policy, using visual media to distribute the
evidence of their administration and engagement with territory. This
occurs when the messages reach domestic audiences, but the value expands
when they reach international audiences.

Reading the semiotics of sovereignty in postage stamps is unique methodology, but with these objects considered as legitimate communications

of the state, 'stamps can and should be read as texts, often with expressly political purposes or agendas which are conveyed through the images they depict'.[61] Through this communication, they perform the legitimisation of their sovereignty in specific bounded spaces as seen in the postage stamps of the US, Canada and Russia. In these states' visualisations of the Arctic, the images showcase their exploration and scientific expeditions, indigenous inhabitants and territory, sometimes accompanied by flags and maps all to the effect of proving their performances of sovereignty in governance and policy of the Arctic.

The body of literature using postage stamps as an empirical data source provides an interesting perspective in showing the performances of government. Themes researched include the political moments, such as transitions in power, nationalism and imperialism. In setting the standard for imperial symbolism in postage stamps, Wallach discusses how Great Britain 'displayed an acute awareness of the power of symbols to shape popular perceptions and to fuel national sentiment, and hence to shape the political reality',[62] setting the standard for other governments to follow in their use of imagery on postage stamps. However, following the practices of reinforcing empire through postage stamps, throughout the twentieth century there has been a significant movement towards independence of territories formerly held by empires and 'with independence comes national rebranding, and among the first actions of any newly independent state is the production of national stamps, notes and coins'.[63] Through postage stamps it is possible to pinpoint this transition and the icons of national significances that became important to the emerging states for reinforcing a common identity. The 'commemorative stamps [that] are issued year after year … reflect most of the events, peoples and places that impinge on a nation's consciousness',[64] and thus postage stamps are a means to evaluate which issues, histories or discourses are important to a particular country and what images are used to construct a national identity. In this way, this book presents some of the symbols of imperialism, nationalism and political transitions including icons and motifs of state authority, and references to territory, effective occupation and the historical and cultural heritage of these countries. It does this by focusing on the structures of political organisation and the policy objectives of the Arctic states by explaining 'social meanings in terms of which it may have been framed'.[65] This establishes that postage stamps have not been issued in a discourse vacuum, and situates them within a specific historical context that reflects both national and international political transitions relating to the development and performance of Arctic sovereignty.

Structure of the book

Chapter 1 will outline in brief the history of the imperial establishment of the Arctic, framed within the timeline of the Colombian-epoch. The chapter will discuss how the rules of imperial conquest, including performances of sovereignty, were used to claim nascent sovereignty over the territory and peoples of the Arctic, drawing boundaries between imperial states. This chapter will explain how the present Arctic sovereignty is situated within the evolution of territorial sovereignty throughout the entire international system. It pinpoints several key historical pivots in normative rules causing the turn away from imperialism to the development of the modern international state system and the impact this had on Arctic sovereignty. This change created the need for states to represent their sovereignty over the Arctic in a new framework of effective occupation, hence the turn to emphasis on government administration. Critically, this is a pivot which required the Arctic states to change tack in their engagement with the Arctic, situated within a rules-based and socially constructed international system when the introduction of new rules created the need for new expressions of performativity.

Chapter 2 relates the history of US engagement with the Arctic over the *longue durée* situated within the narrative of the evolution of US policy in the Arctic region. This telling of the US relationship with the Arctic is developed within the framing of the coded themes of Arctic policy as illustrated in their postage stamps. It utilises first the empirical and primary sources of US policy documents, such as Executive acts, Senate records and Congressional legislative documents as well as commentary from observers of US Arctic policy, including legal scholars. These primary materials are supplemented by the work of existing histories of the Arctic and those who study US Arctic geography and geopolitics. It argues that US postage stamps depicting Arctic themes correlate with policies that implement effective occupation of Alaska through performance of government administration, projecting sovereignty over this territory in accordance with the rules of the international system. It shows that many of their current performances demonstrate the confidence of the US in their claims to Arctic sovereignty.

Chapter 3 narrates the history of Canadian engagement in the Arctic from the empire of Great Britain to the Dominion of Canada through Canadian Arctic policy. It develops a historical narrative of Canadian Arctic sovereignty through Senate records and parliamentary legislative documents. The analysis of these primary documents is supported by the research of Canadian Arctic historians and those who study Canadian Arctic geography and geopolitics. The chapter analyses the Canadian projection of Arctic sovereignty through a case study of its catalogue of postage stamps, arguing that Canadian postage stamps depicting Arctic themes reinforce policy practices that demonstrate effective occupation of the high north, projecting sovereignty over this territory in response to the changing rules of the international system. It shows that many of their current performances reflect the crisis of sovereignty created due to the procedural failures of their initial efforts to extend sovereignty over the Arctic outside of the normative rules for territorial acquisition.

Chapter 4 recounts the history of Russian engagement in the Arctic over the *longue durée* situated within an analysis of Russian Arctic policy preferences as illustrated in Soviet and Russian postage stamps. It develops a historical narrative of Russian Arctic sovereignty and their confidence in their broader Arctic claims, but concern about losing the North Pole as a symbol of Russian mastery and identity. It analyses the Soviet and Russian projection of sovereignty throughout Arctic territories from primary source policy documents and diplomatic statements, supported by historical accounts and foreign policy analyses of the state. It argues that Russian postage stamps depicting Arctic themes reinforce policy practices conveying effective occupation of the vast expanse of the Russian Arctic, promoting legitimisation of their sovereignty over this territory in response to the changing domestic and international structures.

Chapter 5 uses quantitative methods to conduct a comparative analysis of Arctic policy performances of the US, Canada and Russia as depicted in their postage stamps across a range of policy options, and shows the conversion and correlation of domestic policy to foreign policy. The results of this quantitative analysis demonstrate that within each of these states underlying domestic policy trends emerge, revealing the prioritisation of policy agendas of importance to the performance of Arctic sovereignty. For the US, this correlates with a projection as being good environmental stewards of natural resources through images of national parks, flora and fauna. The Canadian priority exists in promoting the indigeneity of their Arctic sovereignty through emphasising the import of indigenous populations into the historical narrative of Canadian state-building, a strategy maximised under the Harper administration. Meanwhile, Russian priorities were emphasised in projections of Russian skill of mastery over nature in their history of polar scientific expeditions and Arctic operational prowess, beginning in Stalinist projects and continuing through to the Putin administration.

The Conclusion brings these three case studies together. Through this merging of policy discussion, it provides the space to discuss how through projections of sovereignty rooted in domestic territorial authority and legitimacy, these states are able to project the aura of sovereignty over the

Arctic. This includes reinforcing legitimisation of their domestic sovereignty over the terra firma of the Arctic, but also in projecting apparent sovereignty over the international spaces of the Arctic, spaces which are not excluded from images of postage stamps by the mere legal technicalities of the delimitation between the domestic and the international. This demonstrates that although the structure of the international system has evolved around these states, away from acceptability of formal projections of imperialism through blatant ceremonies of sovereignty, policy performances are still used to project power and authority over the international spaces of the Arctic. While flag planting has been superseded by more sophisticated forms of performance, states still use policy and the implementation of policy to demonstrate effective occupation of a legitimate authority, and therefore sovereignty over the spaces of the Arctic.

Notes

1 Robert E. Peary, *Diary of Robert E. Peary*, 1909, p. 76.
2 Robert E. Peary, p. 83.
3 Robert E. Peary, *The North Pole: Its Discovery in 1909 Under the Auspices of the Peary Arctic Club* (New York: Frederick A. Stokes Company, 1910), p. 295.
4 Robert E. Peary, 1910, p. 297.
5 Robert E. Peary and R.A. Harris, 'Peary Arctic Club Expedition to the North Pole, 1908–9', *The Geographical Journal* 36, no. 2 (August 1910): 129–44, p. 141.
6 'Taft Has Faith in Peary: Congratulates Him Without Qualifications on Finding the Pole', *New York Times*, 8 September 1909.
7 Mike McDowell and Peter Batson, 'Last of the Firsts: Diving to the Real North Pole' (The Explorers Club, 2007), https://explorers.org/flag_reports/Mike_McDowell_Flag_42_Report.pdf.
8 McDowell and Batson.
9 'The Arctic Triumph of the Chilingarov Expedition', *United Russia*, 3 August 2007, www.edinros.ru/news.html?id=122356.
10 McDowell and Batson, 'Last of the Firsts: Diving to the Real North Pole'.
11 'Russian Flag Planted on N Pole Seabed' (Russia Today, 2 August 2007), www.youtube.com/watch?v=-drWfpNCRb4.
12 'Arctic Expedition Team Return Home After Flag-Planting', *AP Archive* (Associated Press, August 2007), www.youtube.com/watch?v=Xp3twlJJxwY.
13 See Patricia Seed, *Ceremonies of Possession in Europe's Conquest of the New World, 1492–1640* (New York: Cambridge University Press, 1995).
14 Andrew F. Burghardt, 'The Bases of Territorial Claims', *Geographical Review* 62, no. 2 (1973): 225–45.
15 W. Lakhtine, 'Rights Over the Arctic', *The American Journal of International Law* 24, no. 4 (1930): 703–17.
16 See Nicholas Blomley, 'Disentangling Property, Performing Space', in *Performativity, Politics and the Production of Social Space*, eds. Michael R. Glass and Reuben Rose-Redwood (New York: Routledge, 2014).

17 United Nations, 'United Nations Convention on the Law of the Sea – Part VI: Continental Shelf' (United Nations, 1982), www.un.org/depts/los/convention_agreements/texts/unclos/part6.htm.

18 See, for example, US Geological Survey, 'Circum-Arctic Resource Appraisal: Estimates of Undiscovered Oil and Gas North of the Arctic Circle', Science for a Changing World (Menlo Park, California: US Department of the Interior, 2008), https://pubs.usgs.gov/fs/2008/3049/fs2008-3049.pdf.

19 The US have not ratified the 1982 United Nations Convention on the Law of the Sea (UNCLOS) Convention. It is, however, party to earlier versions of the law of the sea. Given the short continental shelf adjacent to Alaska, there is little incentive for the US to increase its continental shelf sovereignty beyond the already existing 200 nautical miles if only to access Arctic resources.

20 See International Boundaries Research Unit, *Status of Arctic Waters Beyond 200 Nautical Miles from Shore*, Arctic Maps (Durham University, 2016), www.dur.ac.uk/resources/ibru/resources/ArcticmapStatusofArcticWatersbeyond200NM.pdf.

21 Alan Freeman and Unnati Ghandi, 'Russian Mini-Subs Plant Flag at North Pole Sea Bed', *The Globe and Mail*, 2 August 2007, www.theglobeandmail.com/technology/science/russian-mini-subs-plant-flag-at-north-pole-sea-bed/article20400041/.

22 John R. Searle, 'How Performatives Work', *Linguistics and Philosophy* 12, no. 5 (1989): 535–58.

23 For more on rule-governed behaviour and speech acts, see John R. Searle, *Speech Acts* (Cambridge: Cambridge University Press, 1969).

24 See, for example, Scott G. Borgerson, 'Arctic Meltdown: The Economic and Security Implications of Global Warming', *Foreign Affairs* 87, no. 2 (2008): 63–77.

25 Klaus Dodds and Mark Nuttall, *The Scramble for the Poles: The Geopolitics of the Arctic and Antarctic* (Cambridge: Polity Press, 2016), p. 4.

26 Lassi Heininen and Matthias Finger, 'The "Global Arctic" as a New Geopolitical Context and Method', *Journal of Borderlands Studies*, 2017, https://doi.org/10.1080/08865655.2017.1315605, p. 1.

27 See, for example, Eyal Weizman, *Hollow Land: Israel's Architecture of Occupation* (London: Verso, 2007).

28 Øyvind Østerud and Geir Hønneland, 'Geopolitics and International Governance in the Arctic', *Arctic Review on Law and Politics* 5, no. 2 (2014): 156–76, p. 156.

29 Alex Jeffrey, *The Improvised State: Sovereignty, Performance and Agency in Dayton Bosnia* (Chichester: Wiley Blackwell, 2013), p. 4.

30 Jeffrey, p. 3.

31 Cynthia Weber, *Simulating Sovereignty: Intervention, The State and Symbolic Exchange* (Cambridge: Cambridge University Press, 1995), p. 4.

32 Phil Deans and Hugo Dobson, 'East Asian Postage Stamps as Socio-Political Artefacts', *East Asia* 22, no. 2 (2005): 3–7, p. 6.

33 Jack Child, *Miniature Messages: The Semiotics and Politics of Latin American Postage Stamps* (Durham, N.C.: Duke University Press, 2008), p. 4.

34 For example, see Stanley D. Brunn, 'Stamps as Messengers of Political Transition', *Geographical Review* 101, no. 1 (2011): 19–36.

35 See Jean Bodin, *On Sovereignty: The Six Bookes of a Commonweale*, trans. M.J. Tooley (London: Basil Blackwell, 1967).

36 See Stephen D. Krasner, 'Abiding Sovereignty', *International Political Science Review* 22, no. 3 (2001): 229–51; Cynthia Weber, *Simulating Sovereignty: Intervention, The State and Symbolic Exchange*; Winston P. Nagan and Aitza M. Haddad, 'Sovereignty in Theory and Practice', *San Diego International Law Journal* 13 (2012): 429–520.

37 See the social contract theories of Hobbes, Locke, Rousseau....

38 Michael Billig, *Banal Nationalism* (London: Sage Publications, 1995).

39 Allen Buchanan, *Justice, Legitimacy, and Self-Determination: Moral Foundations for International Law* (Oxford: Oxford University Press, 2004).

40 See A.R. Clute, 'The Ownership of the North Pole', *Canadian Bar Review* 5, no. 1 (1927): 19–26.

41 See Hans Aufricht, 'On Relative Sovereignty', *Cornell Law Quarterly* 30 (May 1944): 318–49.

42 Patrick Macklem, *The Sovereignty of Human Rights* (Oxford: Oxford University Press, 2015), p. 33.

43 James Anderson and Liam O'Dowd, 'Borders, Border Regions and Territoriality: Contradictory Meanings, Changing Significance', *Regional Studies* 33, no. 7 (1999): 593–604, p. 594.

44 See Jon D. Carlson *et al.*, 'Scramble for the Arctic: Layered Sovereignty, UNCLOS, and Competing Maritime Territorial Claims', *SAIS Review of International Affairs* 33, no. 2 (2013): 21–43; Corine Wood-Donnelly, 'Constructing Arctic Sovereignty: Rules, Policy & Governance 1494–2013' (Brunel University London, 2014), Brunel University Research Archive, http://bura.brunel.ac.uk/bitstream/2438/10542/1/FulltextThesis.pdf.

45 See, for example, Judith Butler, 'Performative Acts and Gender Constitution: An Essay in Phenomenology and Feminist Theory', *Theatre Journal* 40, no. 4 (1988): 519–31.

46 See Judith Butler, 'Giving an Account of Oneself', *Diacritics* 31, no. 4 (2001): 22–40.

47 Mark B. Salter, 'When the Exception Becomes the Rule: Borders, Sovereignty, and Citizenship', *Citizenship Studies* 12, no. 4 (2008): 365–80, p. 366.

48 Wendy Brown, *Walled States, Waning Sovereignty* (Brooklyn, NY: Zone Books, 2010), p. 26.

49 Eyal Weizman, *Hollow Land: Israel's Architecture of Occupation*, p. 7.

50 Rachel Busbridge, 'Performing Colonial Sovereignty and the Israeli "Separation" Wall', *Social Identities* 19, no. 5 (2013): 653–69, p. 659.

51 Elke Krahmann, 'Legitimizing Private Actors in Global Governance: From Performance to Performativity', *Politics & Governance* 5, no. 1 (2017): 54–62, p. 54.

52 Jeffrey, *The Improvised State: Sovereignty, Performance and Agency in Dayton Bosnia*, p. 2.

53 Cynthia Weber, *Simulating Sovereignty: Intervention, The State and Symbolic Exchange*, p. 127.

54 Michael R. Glass and Reuben Rose-Redwood, eds., *Performativity, Politics, and the Production of Social Space* (New York, New York: Routledge, 2014), p. 3.

55 John R. Searle, 'How Performatives Work', p. 537.

56 J.L. Austin, *How To Do Things With Words* (Oxford: Clarendon Press, 1962), p. 6.

57 Nicholas Onuf, 'Speaking of Policy', in *Foreign Policy in a Constructed World*, ed. Vendulka Kubálková (London: Routledge, 2015), p. 86.

58 Stanley D. Brunn, 'Stamps as Messengers of Political Transition', p. 19.

59 See A. Ayalon, 'The Hashemites, T.E. Lawrence and the Postage Stamps of the Hijaz', in *The Hashemites in the Modern Arab World*, eds. A. Susser and A. Shmuelevitz (Abingdon, UK: Frank Cass & Co, Ltd, 1995), 15–30.
60 Corine Wood-Donnelly, 'Messages on Arctic Policy: Effective Occupation in the Postage Stamps of the United States, Canada and Russia', *Geographical Review* 107, no. 1 (2017): 236–57, https://doi.org/DOI: 10.1111/j.1931-0846.2016.12198.x, p. 240.
61 Phil Deans and Hugo Dobson, 'East Asian Postage Stamps as Socio-Political Artefacts', p. 4.
62 Yair Wallach, 'Creating a Country Through Currency and Stamps: State Symbols and Nation-Building in British-Ruled Palestine', *Nations and Nationalism* 17, no. 1 (2011): 129–47, p. 138.
63 Keith Jeffery, 'Crown, Communication and the Colonial Post: Stamps, the Monarchy and the British Empire', *The Journal of Imperial and Commonwealth History* 34, no. 1 (2006): 45–70, p. 63.
64 Igor Cusack, 'Tiny Transmitters of Nationalist and Colonial Ideology: The Postage Stamps of Portugal and Its Empire', *Nations and Nationalism* 11, no. 4 (2005): 591–612, p. 593.
65 Ingmar von Homeyer, 'Symbolic Policies and "Rational" National Interests: Explaining the Crisis of European Integration in 1965/66', in *Political Symbols, Symbolic Politics: European Identities in Transformation*, ed. Ulf Hedetoft (Aldershot: Ashgate, 1998), p. 148.

1 Who owns the Arctic?

Introduction

The Arctic region has been the object of geopolitical gaze during various cycles over the last several centuries, periods of engagement that included imperial expansion, scientific exploration and Cold War insecurities. Today, the Arctic is again a global focal point and there are several drivers behind the recent revival of interest. Some of these drivers include factors such as climate change reducing the ice cover, the availability of technology better suited to withstanding the harsh northern climes, and growing economic interests in the resources and development possibilities of the region. Taken together, these interests position the issue of Arctic sovereignty at the centre of intense geopolitical speculation. One of the biggest myths in this flurry of interest is the idea that the Arctic region is at the centre of a scramble for territorial claims. While the idea of an Arctic land grab is pervasive in popular discourse, perhaps mirroring the discourse in past cycles of interest in the Arctic, the present situation is rather different, as 'anarchy does not reign at the top of the world; in fact, it's governed in a manner not unlike the rest of the planet',[1] a position held by much of the academic community. It is the purpose of this chapter to address the often-repeated query 'Who Owns the Arctic?' against the backdrop of the cycles of Arctic fervour showing sovereignty of the Arctic in different temporal positions.

The Arctic is governed according to the norms and rules of territorial sovereignty widely understood throughout the international system. Although governance of the region follows global standards, there are some grounds for the conditions of uncertainty that prevail in discussions of Arctic sovereignty. As it happens, both our understanding of the geography of the Arctic and the rules that frame the game of sovereignty for the region have changed significantly during the long twentieth century. This is most evident by these 'new' claims to Arctic territory, seen in the current filing of claims for extended delimitation of the continental shelf, one of the actions underpinning the appearance of an Arctic scramble in contemporary geopolitics. Beyond this, of some significance in this discussion is that man's knowledge about this Polar Region and its incorporation into geographical knowledge has had a lengthy evolution. In brief, let us first consider the point that our understanding of the geography of the Arctic, and especially that of the North Pole, has changed.

The object in many claims to sovereignty and the focus of exploration for much of the last century has been fixated on a small geographical point in the centre of the Arctic: the North Pole. What is extraordinary about this fact is that the celestial axis of the earth at the North Pole is an entirely imaginary construction. Even though it is not a place that can be reached with any near certainty except by the calculations of scientific instruments, the North Pole has a symbolic status in geopolitical imaginations. With both the North and South poles designated as the celestial centres of the earth, they gained an almost metaphysical sacredness. It has been said that 'these are the spots pierced by the axis of the heavens; they are the crowns of the world, about which all the stars dance, the points to which all compasses direct their needles'.[2] Through this, somehow the North Pole and the surrounding Arctic region have gained a cosmology related to its geographic characteristics and its ideational quality as an object, or a material thing to be possessed.

The idea of the North Pole as a geographical place has its origins in the introduction of the spherical earth concept in ancient Greece; it is Pytheas who is credited with being the first geographer and explorer to discover the Arctic region.[3] He returned from his journeys with reports on 'the sleeping-place of the sun' and gave descriptions of materials that most certainly would have been broken-up pack ice.[4] Later, Mercator is said to have drawn his map of the Polar Regions from a collection of knowledge assembled since Pytheas' voyages, a map that appears to incorporate mythological beliefs from northern indigenous peoples.[5] It is perhaps these cosmological ideas about world rivers that 'flows into the icy seas of the north … [where] It contains an Island of the dead' that Mercator used to depict his inward drawing rivers around four magnetic islands at the North Pole – information that would later go on to feed the long held beliefs about the Open Polar Sea and the existence of another continent at the North Pole.[6] These maps would inform those who would explore the Arctic region: Frobisher, Davis and Hudson, adventurers whose exploration gave credibility to the claim that Queen Elizabeth held legitimate ownership of territories to the North Pole.[7]

However, this was not the first, nor the last claim to ownership of the North Pole and the space that surrounds it. The answer to 'Who Owns the

Arctic?' is not as simple as compiling a short list of those who have made claims to this territory throughout history. The answer ultimately requires a complex explanation of how sovereignty has evolved through structural power in the modernisation of the international system, as well as an understanding of the different layers of sovereignty found within the law of the sea. It is a response that traces not only the history of claims to the Arctic and the North Pole, but also presents a genealogy of the rules, sometimes given as international principles or laws, that provided the glue to legitimise these claims at certain points in history. Understanding these processes is important, as not only does it show who has 'owned' the Arctic at various points in history, but it also helps us to understand who 'owns' the Arctic today, and to consider where 'ownership' of the Arctic may be going in the future.

Like any good mystery, the case of sovereignty and the melting Arctic requires a methodical process to review the facts and evidence of the situation. The method used to understand how the sovereignty of the Arctic came to be situated in its contemporary context is the procedure of process-tracing, a method that 'trace[s] backward the causal process that produces the case outcome, at each stage inferring from the context what caused each cause … lead[ing] the investigator back to a prime cause'.[8] Beginning with the present circumstances of sovereignty in the Arctic, one can trace backwards throughout history to establish who, when, where, why and how the Arctic came to be owned by the littoral states, by considering the steps that changed the rules of sovereignty throughout history. In the process of tracing back through the case of the melting Arctic, this chapter discusses what it means to own the Arctic through the concept of sovereignty and gives snapshots of different conceptions of Arctic sovereignty at different points in history. Here, the case of 'Who Owns the Arctic?' is presented in a chronological narrative, with each stage presented as a cross-section of Arctic sovereignty across a range of temporal and spatial contexts.

Defining ownership: sovereignty

When 'Who owns the Arctic?' is asked, what is really meant is 'Who holds the legitimate authority over decision-making?' or, who has sovereignty for this territory. A second question that underpins this query is the issue of who holds the property rights to the Arctic and thus the benefits to exploit the resources that lie within this territory. Although within political studies, sovereignty is ordinarily conceived of as authority rather than as property rights, it is yet true that states hold the ultimate decision-making authority over that land, as well as owning the property or exploitation rights to land as well. Here are some examples of how some of these nuances between ownership and authority play out in Arctic spaces.

On a circumpolar map of the Arctic, most of the region is coloured into some form of political organisation indicating the political authority of that territory. What these maps do not show is that within Alaska, the US government is the largest landowner[9] and the federal government holds the ultimate decision on what development takes place within this territory, including the issuing of licences for oil or mineral exploration. In Canada, although it has recently settled many land claims agreements along its northern borders with indigenous groups, the boundaries for those land claims extend only to the territorial seas and do not include the continental shelf or its resources.[10] Under Soviet administration, land in Russia was amalgamated through the policy of collectivisation and placed under state control, and in the post-Cold War period, private land ownership continues to be a murky area under Russian law where indigenous people are required to lease land for traditional economic activities from local municipalities. Here, then, are three states with vested interests in the sovereignty of the Arctic due to the implications for their national interests, each as private landowners[11] and each being sovereign and holding sovereignty over this territorially stratified region.

Sovereignty has a basic definition of the 'supreme power or authority',[12] and it is 'an idea of authority embodied in those bordered territorial organizations we refer to as "states …".[13] It is a complex term with a much-deliberated significance, varying usage and applied contradictions, as well as disciplinary contestations in its usage. Within this tangle of discourse, it has been described as 'organised hypocrisy'[14] and despite the reliance of governments that wield the term, it is argued to be a disintegrating method for describing the authority of the bordered state.[15] However, the term often appears in Arctic discourse, referring to both domestic and international policy aspects of Arctic states, and this induces the need to unpack what these states mean when they refer to 'Arctic sovereignty'. It is particularly perplexing when diplomats and government officials make statements with reference to Arctic sovereignty using verbs such as 'to establish', 'to exercise', 'to maintain'; how is it that after centuries of exploration, commercial exploitation and decades of scientific investigations, the sovereignty of the Arctic is still in question? There are several aspects to this conundrum.

There are two dimensions to the condition of being sovereign and it is achieved when a state holds both internal and external sovereignty.[16] Internal sovereignty is when within a state 'there are none who acknowledge any other sovereign'.[17] Emerging in medieval Europe, this conception of sovereignty becomes *raison d'état* – the reason of state, where the 'acquisition, maintenance and expansion of sovereign power'[18] is an attribute of a monarch who is accountable to no higher authority. Predating the Westphalian system of sovereign states, this concept of sovereignty developed when the Pope wielded significant political and religious authority; perceived as the messenger of God, he served as arbitrator between warring monarchs. The decline of the monarch as the sovereign authority, the waning power of the Pope and the consolidation of principalities occurring in this period introduced change for the fledgling international system, including changes in the concept of sovereign authority.

Corresponding with the decline of papal authority, the concept of sovereignty advanced, becoming defined as the 'absolute and perpetual power of a commonwealth'[19] where sovereignty was invested not in the monarch, but in the ruler's obligations to the citizens on the basis of contractual agreement. In this period of political enlightenment, the absolute authority of the monarch was under attack when the rights of the people directly challenged it. This challenge emerged in the social contract theories of Hobbes, Locke and Rousseau, who for varying reasons, including fear and security of property, determined that individuals choose to surrender the freedoms accorded them by the laws of nature in the hypothetical state of nature to an authority figure – a sovereign. As a result, the legitimacy of the sovereignty of the state becomes inextricably linked to the consent of the governed and through the evolution of parliamentary representation, the monarch's sovereignty is consolidated with the internal sovereignty of the state.

Yet the advancement of the joint investment of sovereignty between the ruler and the ruled becomes problematic when Europeans transplant the norm of sovereignty to the New World. When this happens, the imposition of authority outside of the social contract again becomes a reality when European monarchs impose their authority over non-consenting indigenous subjects. Sovereignty over newly discovered lands and inhabitants was rooted in at least two different principles, one religious, the other material. An early justification of the Iberian empires, granted their entitlements by the Pope, was based in ecclesiastical obligation, in the notion of the universal sovereignty of God and 'the obligation to bring all infidels and pagans into a state of Christianity and, therefore, civility'.[20] However, as the sovereignty of states became joined with consent and as the notion of territorial integrity entered into the fledgling international system, new forms of justification were required for annexing new territories, especially with the decline of papal authority.

To challenge the 'first discovery' claims of the Iberian powers and to denigrate their position of religious right, the rest of Europe needed new methods to adapt the norms of territorial sovereignty for the New World in their favour. This came in the material justifications for the appropriation and conversion of common land into private property, explained through the idea that although God may have given the earth to all men, this bequest came with a caveat. This was presented in

> a justification of appropriation which would do double duty, both in England and in America ... 'God gave the World to Men in Common; but ... it cannot be supposed he meant it should always remain common and uncultivated. He gave it to the use of the Industrious and Rational.'[21]

The result is the *labour theory of property*, which suggested that any land wasted through lack of cultivation was available for appropriation. Through this manipulation, European powers determined that New World lands were underutilised by the inhabitants, as well as by the Iberian empires, and were available for new territorial claims beyond the efforts of religious conversion and overriding the prior land-use and habitation of indigenous populations.

Following this designation of self-entitlement to the New World, the:

> European nation-states proceeded to colonize North America by making grandiose territorial claims on the basis of discovery, papal bulls, symbolic acts of possession, royal charters, and settlement, as though the continent was juridically vacant and the Indigenous peoples living there did not have sovereignty.[22]

Although it was possible to rationalise the appropriation of unused land through the notion of added labour, there is one aspect of this process that proved problematic: the sovereign status of indigenous inhabitants. This sovereign status and the rights of indigenous peoples was explained away by exposing the differences in political organisation, by justifying one system over another and enforcing it through asymmetric power. Europeans determined that because indigenous people lacked a recognised universal and supreme authority, they could not be considered internally sovereign.

By applying their own legal codes and definitions, European civilisations determined that indigenous societies in the New World lacked internal sovereignty. As a result, they also denied them the status of external sovereignty and equality with other political units in the international system. External, or international sovereignty, 'is defined by three characteristics: territory, autonomy, equality'.[23] The condition of sovereignty implicitly established in the *Peace of Westphalia 1648*,[24] sometimes

considered to be the starting point of the modern international system, pro-
liferated the rule of external sovereignty in international relations.
However, even if indigenous peoples had been considered to hold internal
sovereignty, political thought at the time also permitted for the 'rights of
other sovereigns to enforce justice in the entire world';[25] a ruler could be
deemed to have in some way violated the social contract with its subjects
and therefore be deposed.

Because the concept of sovereignty described the authoritative relationship between the ruler and the ruled, there was a direct correlation to the
rights of a monarch to levy taxation on his subjects, so in the European
context there was an 'inseparable connection between land tenure ... [and]
genuine sovereignty'.[26] As the idea of sovereignty over a fixed territory
with its population and the use of cartography to represent territorial possessions became normalised in the international realm, states encountered
two problems. The first is that they experimented with the possibility of
maintaining continuous sovereignty from one land to another over the
ocean in debates between *mare clausum* and *mare liberum*.[27] Second, as
they made sovereignty claims to sparsely populated territories with the
land takeovers legitimised through the labour theory of property, the lack
of subjects over which to exercise authority required the generation of new
methods for exercising the administration of sovereignty, a practice and
performance that came to be known as effective occupation.

The debates between the closed seas, *mare clause*, and the free seas,
mare liberum, was an issue that ultimately determined whether it was possible to exercise sovereignty over the oceans in the same way that it was
exercised over land. The contention between the two approaches is rooted
in the practices of some Mediterranean city-states who claimed their near
waters as sovereign territory[28] contrasted with legal practices of Romans,
who 'held that all the peoples of the world possessed a universal right to
travel and trade'.[29] At first, Spain and Portugal claimed authority over large
swathes of ocean, creating monopoly conditions on trade routes between
Europe and its imperial territories. The rest of Europe challenged this position through the legal arguments of Hugo Grotius who posited these claims
as 'far fetched and unjust'[30] as their techniques could not constitute rivalry
or even fair competition. However, the idea of enclosed seas soon had a
revival when John Selden argued that the British had a right to maritime
dominion because their 'sovereignty in territorial waters was based on long
and continuous possession'.[31] In many ways, the debate between *mare
clausum* and *mare liberum* was a contest between resource materiality
versus the commons, and this debate yet continues.

These opposing approaches found a middle position in a three-mile rule
for territorial waters, a standard maintained until the mid-twentieth century

when the contest between the two approaches was reignited by a challenge to the norms of maritime territory. These challenges came in the form of a unilateral declaration by the US to extend their sovereign rights over the submarine continental shelf for exploitation purposes, an action which quickly gained international consensus. Since then, the law of the sea has been codified with new forms of maritime territory including extending the limits of the territorial waters, and introducing the idea of the continental shelf and exclusive economic zone. As we enter a new age of awareness of plastics pollution and climate change, it may be that there are yet new normative changes to expect with regards to ocean governance and the responsibilities of states within delimited areas to correspond with the rights of exploitation that they claim.

The second problem that imperial states encountered involved the paradox between sovereignty as derived from the social contract in juxtaposition with the use of the labour theory of property to justify claims to territorial annexation. The normative rules for extending sovereignty over new territory began with discovery and symbolic performance of ceremonies of possession, but that was 'capable of granting only an inchoate title which must in due time be completed by actual occupation'.[32] Not only did European sovereigns lack a mutually constituted political relationship with indigenous persons, but they were also failing to use the land through the application of labour to fulfil the legal requirements for effective occupation.

Effective occupation was ultimately fulfilled in the New World through the development of industrial land-use as well as settling the land through colonisation. In the Arctic, however, the stringent requirements for effective occupation were waved in lieu of government administration meeting 'reasonable requirements'.[33] It is this exception that led to the creative mechanisms used by Arctic states to demonstrate their effective occupation over the region, including scientific exploration, environmental stewardship, residential schools and resource exploitation.

Despite this creativity, throughout much of the Arctic the condition of 'perfect' sovereignty, both internal and external, remained elusive. This is true both for states making claims to sovereignty and for the status of indigenous inhabitants living in the region. This imperfect condition has been perpetuated by changes to the climate, the increasing accessibility of the Arctic, changes to the rules for annexing new territory and finally, changes in the materiality of the oceans as modifications to the law of the sea increasingly placed the oceans into a framework of delimited sovereign spaces. These spaces represent a compromise in the debate between *mare liberum* and *mare clausum* by maintaining the principle of free navigation while designating the resources of the oceans as a material good. The effect has been to reduce the oceans as a commons and to expand the spaces that fall under the state's national realm.

Designating 'who owns the Arctic', sovereignty is a rule that establishes normative order by establishing political authority and rights to economic exploitation in specific spaces. The condition of sovereignty 'represents claims of rights, specifically the right to rule and the right of authority. As a property right, "sovereignty" is the highest, most complete right of ownership.'[34] The Arctic states are following the norms of sovereignty; however, these rules have seen significant evolution over the century. The myth that the Arctic is at the centre of a territory scramble is perpetuated by the lack of perfect sovereignty in the region, a condition created by the changing rules in combination with the communication of state's practices and performances of sovereignty for demonstrating effective occupation of the region.

There are eight states with territory in the Arctic or sub-Arctic, and five of these also border the Arctic Ocean. At present, the land boundaries of the Arctic states are stable, but within the maritime spaces, significant expansion is underway given the recent introduction of new forms and parameters for the maximum extent of territory in the law of the sea, including the exclusive economic zone and the continental shelf. Current claims to sovereignty in the Arctic, although introducing the movement of territorial boundaries (a situation that usually brings conflict and insecurity), are in fact reproducing stable normative understandings. Through the types of performances states employ to demonstrate their claims to Arctic territory, they are using known and recognised rules of spatial annexation and they are repeating these performances of territorial authority to broadcast and confirm their legitimate sovereignty over the Arctic. Yet how did the current rules for territory come to frame the Arctic as it is today?

What now follows is a discussion of who has owned the Arctic in different combinations of rules ever since the discovery of the New World.

Arctic sovereignty in the Age of Discovery: 1492–1908

Setting the stage: the rules of territorial expansion

The movement of European explorers searching for a new route to Asia and the Indies by sailing west set in motion the development of laws regulating the processes of territorial expansion. Columbus never truly discovered the New World, as 'actually millions of human beings were already living full and imaginative lives on the continent in 1492. That was simply the year in which the sea pirates began to cheat and rob and kill them.'[35] The implication of discovery was to determine which European was first to sight a new territory, the first step in the developing rules of procedure for the annexation of new and formerly 'undiscovered' territory by European empires. The traditions established during this period developed into customary international law, providing the still existing standards for maintaining territorial sovereignty. However, almost all lands, discovered or undiscovered, are in some way claimed by contemporary states.

Rules 1 and 2: discovery and ceremonies of possession

During this period, new land was often claimed by explorers on official missions for European sovereigns intent on expanding state territory for access to resources or sea routes. Discovery of new territory gave right to first possession. Given that explorers often used the cartographic knowledges of previous adventurers, it was necessary to leave a trace for those who might eventually reach the same territory to indicate the land had already been discovered. For this reason, explorers left symbolic mementos to inform others that the land was no longer free. These symbolic acts 'may be interpreted also as a device to show to the world that an inchoate title to the discovered region was acquired which rendered it *terra prohibita* as far as other States were concerned'.[36] It is the use of these symbolic acts in contemporary Arctic international politics that has often caused great consternation, even though they no longer hold significance within international law.

Symbolic acts varied greatly in practice from state to state. The Spanish were fond of unfurling flags while Russian practice included the burial of copper plates with the inscription *Russia Imperial Territory* over which a cross was erected. In the case of Alaska, Behring 'left a number of articles, consisting of beads, an iron kettle, and some coins, as evidence of his presence there', and this was sufficient evidence of Russian ownership when Cook explored the region in 1778.[37] Peary understood the necessity of symbolic acts accompanying claim to discovery, and upon his arrival at the North Pole in 1909 also executed symbolic acts to indicate that he was first to acquire the pole. The importance of symbolic acts in claiming territory and its universal recognition as the first step to sovereign control of an area is intriguing given the performance of flag planting on the seabed at the North Pole and the discourse of 'firsts' that accompanied the subsequent press releases.

However, the rules of the international system developed so that it was necessary for symbolic annexation to be followed by effective occupation. Discovery and symbolic acts granted only initial rights and a rudimentary,

albeit worthless title that 'finally perishes unless it is followed and per-
fected by effective possession in a reasonable time'.[38] The advantages of
first discovery and ceremonies of possession had to be followed eventually
by use of the territory and extension of government administration.

Rule 3: effective occupation

Discovery and symbolic acts were established in international custom as the
first elements of claiming sovereignty over new lands, but these had to be
followed by performance of the third rule: effective occupation. This rule of
effective control 'implied continuous administration and effective occupying
of the land; ideally, the territory should be settled throughout and the natural
resources of the area should be developed and used'.[39] Historically this was
demonstrated in the New World through the establishment of trading companies – such as the Hudson Bay Trading Company, establishment of territorial governors and the placement of colonies, granted existence through
charters issued by European monarchs.

However, the notion of effectiveness has variable application, as 'the
nature of the territory affects the degree of sovereign activity needed to
establish effective control'.[40] In the Arctic, the harsh climate is generally
unsuited to settler colonialism, and when the fur trade declined as a
method of government administration, the sparse indigenous populations
were reorganised to provide a paltry substitute for evidence of government
control. In the Arctic, effective occupation came to be understood in the
terms of 'as much as could be reasonably required' given the climatic conditions, and instead states began to operate on the notion of regions of
attraction.[41]

The remainder of the Arctic region is assorted ocean spaces, which currently fall under the jurisdictional areas of territorial waters, exclusive economic zones and the high seas regimes. In the contemporary Arctic,
demonstration of effective control and occupation still finds its way into
the language of policy documents and diplomatic statements in the style of
'maintaining' or 'establishing' sovereignty. Today, states perform this
effective occupation through a variety of political technologies, many of
which are visually recorded in the tiniest messenger of the government, the
postage stamp.

Act 1: the game begins

The year 1492 is arguably one of the most important moments in both the
development of the Arctic and of the international system, as it is the year
that sparked a rapid acceleration of the expansion of European knowledge.

When Columbus sailed the ocean blue 'discovering' the New World, the processes of the Columbian exchange began with the expansion of European empires culminating in an irreversible global transformation. While one does not usually associate the adventures of Columbus with the Arctic, his actions and the political events of his time provide the seed from which germinated the development of the international legal order relating to maritime and territorial law which now frames the whole narrative of the territoriality of the Arctic. School children learning the modern history of Western civilization are given this date as a starting point of history; it is also the starting point for the development of Arctic sovereignty.

Although Europe had long since moved beyond the idea of a flat earth, maps used by navigators of the fifteenth century still depicted a world of unknowns. This includes a map drawn by Columbus circa 1490 which shows the known discoveries of Portugal in Africa,[42] but nothing was known beyond the Atlantic Ocean. As 'Portuguese expansion along the coasts of Africa had established a sequence for gaining possession of these lands: first, discovery; second, construction of a fort and establishment of a garrison; third, a papal bull granting possession to Portugal',[43] these Portuguese discoveries provided the impetus for Spain to also secure new resources. And while seeking an alternative route to India, Columbus discovered the New World for Spain.[44] Here they repeated the established procedures and performances to establish jurisdiction over their new territories.[45]

This pivotal year was followed by two eventful years of exploring, discovering new territory and making territorial claims for the Iberian empires, leading to confusion and conflict over which power owned what territory. Based upon recommendations from Columbus,[46] the Pope divided the Western hemisphere between Spain and Portugal 'in virtue of his territorial supremacy over the whole world'.[47] Spain received the territory to the west of the division line and Portugal, the territory to the east. Even though this meant a diminished role in the New World, Portugal accepted the jurisdiction of the Pope, the de facto head of the international system as this was also the source of their supremacy in Africa.[48] Yet within two years, these powers had determined that the terms of the delimitation were inconvenient, both in its placement in the Western hemisphere and in its absence in the Eastern hemisphere.

In the now seemingly insignificant Spanish village of Tordesillas, Spain and Portugal negotiated 'possibly the most pregnant treaty in world history'.[49] Here, they bypassed papal authority and changed the placement of the line of demarcation.[50] The treaty states 'that a boundary or straight line be determined and drawn north and south, from pole to pole, on the said ocean sea, from the Arctic to the Antarctic pole'.[51] An idea often forgotten in the discourse of Arctic international relations is that in 1494, as *terra nullius* and otherwise lacking the complete formal incorporation of territory within the political boundaries of states, the Arctic belonged to Spain and Portugal with the line of territorial demarcation placed along the division of the hemispheres, drawn from pole to pole. Yet Spain and Portugal failed to discover or symbolically fulfil any of these accepted rules of territorial conquest. As a result, the territory bequeathed through a declining form of legitimacy and their inchoate titles failed when they did not effectively occupy the Arctic, leaving it open for future imperial conquest.

Act 2: new players and new rules

Around the middle to late sixteenth century, there would be a shift in international dynamics that would change and expand the sovereignty of the Arctic in response to the addition of new normative rules for the international system. From Western Europe, upcoming powers challenged Spain's ownership over the territories in the Western hemisphere. From Eastern Europe, exploration and sovereignty would expand around the globe until all the North American Arctic was claimed as the sovereign territory of several European states: Great Britain, Denmark and Russia. This process would introduce two of the major principles of international law still in practice in today's Arctic and all the maritime spaces of the globe: the freedom of the seas and territorial waters.

This process began in a direct challenge to the *Treaty of Tordesillas* in an arrangement known as lines of amity, where Iberian authority was acknowledged up to a certain geographic point, not far from the European continent, but after this point there was no peace beyond the line.[52] The arrangement resulted in the weakening of the extended Spanish authority over the vast expanses of the New World and their ability to maintain this vast empire became compromised, additionally complicated by policies nearer to home, including oppressive fiscal and religious policies against their other European subjects. When Spain failed both to suppress these challenges and to modernise with the economic shift to mercantilism, their global authority was challenged by power parity and their declining fortunes created the opportunity for new players and new rules of the game.

One of the challenges to this Iberian supremacy emerged when Great Britain and the Netherlands established trading companies at the beginning of the seventeenth century. These privateering initiatives were in practice used at sea to challenge Portuguese dominance over trade routes to the East, proving their ineffectual ability to control the sea over which they claimed sovereignty. This turn to trading of resources from exotic places created a source of revenue from outside the state, paid for ships and crew and increasing sea power. This decreased the relative strength of the Iberian powers, while making friends and allies through trading partnerships.[53] Once this authority dissolved, trading companies were then used to demonstrate effective occupation and sovereignty on land in the North American Arctic with the Russian-American Company in Alaska and the Hudson Bay Company in Canada.

In addition to this performative challenge, a legal approach contested the authority of the Iberian powers over the earth's oceans. Grotius, a legal scholar working for the Dutch East India Trading Company, wrote a treatise to convince Europe of the rights of the Netherlands to access the wealth of world trade that Spain and Portugal had monopolised. His defence of their right to global trade routes appears in *Mare Liberum* or 'the free seas' where he contends the Iberian assertion of control over all global maritime territory is 'far fetched and unjust'.[54] Grotius made evident that the purpose of his argument was to defend the interests of the trading company, the Netherlands and the whole of the human race, invoking the laws of nature and the laws of territorial possession. Through this, Grotius systematically argues point by point why Spain and Portugal have no right to prohibit other European nations from expanding their trade networks, demonstrating they have failed in effective occupation. He succeeds, and from this point in history, the freedom of the seas has been maintained as a dominant principle of international maritime law, preserving the world's oceans as a global maritime highway.

The new rule found in the principle of the freedom of the sea facilitated the rapid expansion of global trade and a revolution of the international system. This includes many of the developments in international law and commerce that would shape international politics for next several centuries through 'the establishment of a new and qualitatively different type of international system, based not only on military prowess, but also on industrial strength and commercial superiority, especially in banking, shipping and shipbuilding'.[55] Yet alongside this moment, there was another struggle to determine whether the seas truly were the common property of all and the private property of no state as territorial and power rivalries caused states to compete for control of the sea for its resources.

This next challenge presented itself in the practices of Great Britain in the early seventeenth century, in domestic policy 'that fishing in English waters should be reserved for English vessels'.[56] This had the effect of severely limiting the fish stocks available to European fishermen on the continent while allowing English fishermen better access to fisheries and trading. This policy prevailed because England had the superior navy to support this policy. In defence of this policy of protecting 'English' seas and as a response to Grotius' *Mare Liberum*, Selden penned *Mare*

Clausum arguing 'that the sea could be appropriated through law and custom [and that] English sovereignty in territorial waters was based on long and continuous possession'.[57] This policy was predominantly to protect the interests of its fishermen in the waters surrounding the British Isles, but they never attempted 'the prohibition of free navigation on the so-called British Seas'.[58] By making this claim, England introduced the principle for what would become known as *territorial waters*, a maritime zone where states enjoy exclusive rights of resource exploitation. This performance of territorial authority, reinforced by military power, created the provision within international law for a new rule of territorial waters and positioned Britain as the great naval power, propelling its expansion into the New World.

Great Britain's territorial empire extended to the North American Arctic through discovery and claims to the Hudson Bay watershed. To manage effective occupation over this territory, an estimated 15 per cent of land in North America,[59] they established the Hudson Bay Company in 1670 as a 'body corporate and politic forever', giving the company powers normally associated with the state. The company performed effective occupation by trapping furs for European markets until continental conflicts between Britain and France found their way into the New World. Following a temporary loss of territory to France, through the leverage of the British Crown, the Hudson Bay Company reclaimed their possessions in the *Peace of Utrecht 1713*. This treaty reaffirms the customs of international law regarding the acquisition of territory as Britain succeeded in placing the word 'restore' in the terms of the treaty. In this, France acknowledged that it had overstepped its economic activity in the region. Over the next century, the British Crown would establish a representative government in Canada and created the Dominion of Canada as a political entity in 1867, eventually transferring its Arctic lands and islands to Canada in 1880.

While European powers were discovering, exploring and occupying the Western hemisphere around the Atlantic, Imperial Russia was annexing Siberia and the Russian Czar Peter the Great was 'curious to know if Asia and America were separated by the sea, or if they constituted one undivided body with different names'.[60] Heading eastward from Europe, the North American Arctic would also be explored, possessed and incorporated cartographically into the extended imperial possessions of Russia. Imperial Russia first expanded across the vast expanses of Siberia before improving its naval competencies to explore and map the Bering Sea and Russian Alaska in the mid-eighteenth century. On behalf of the Russian government, Vitus Behring discovered the Bering Strait, which connects the Pacific Ocean with the Arctic Ocean. This expedition, followed by several others, gave 'the title of Russia to all these possessions ... derived from prior discovery which is the admitted title by which all European Powers have held in North and South America....'[61]

The development and performance of cartographic practices had the function 'in serving as a record of territory over which the Russian state could demonstrate discovery and administrative history and in also serving as a record of the locations and availability of commercially viable resources',[62] and as a result also served as performances of sovereignty. In order to maintain effective occupation of their North American possessions, the Russian-American Company received a charter from the Russian government to carry out this duty by proxy, granted all rights and duties associated with sovereignty in a territory under the laws of territorial acquisition: exclusive privileges of economic exploitation, the authority to claim new territories in the name of the motherland, to establish and defend settlements and to instigate the development of trading networks with neighbouring countries.[63]

While the Russian-American Company was used to demonstrate the effective occupation of Russia over Alaska, it was being threatened by pressures from other powers from the East and the South. Russia saw Alaska as 'an investment rather than a colony and had vested all her interests in the Russian-American Company'[64] and consequently was without infrastructure of permanent defence or well-established colonies. The result is that Russia was afraid of losing Alaska. Ultimately faced with imperial overstretch and corruption within the Russian-American Trading Company, Russia was motivated to sell Alaska rather than face losing their investment due to a lack of effective occupation. From the European to the Asian continent, Arctic space has been a consistent part of the sovereign domain of first Imperial Russia, the Soviet Union and now, the Russian Federation.

Act 3: introducing the new Arctic empires

The year 1867 was a pivotal date for the evolution of Arctic sovereignty as there was a dramatic shift in ownership of North American territory. Due to influences of the doctrine of Manifest Destiny, the US had interests in expanding its territorial possessions and the potential purchase of Alaska was discussed with Russian diplomatic staff in Washington. Eventually, a Russian minister was sent to the US capital 'with instructions to sell the territory. More significant than the motive of friendship was the fact that this possession had become of no value to Russia and was a burden to the government.'[65] And so the US purchased Alaska with the *Alaska Purchase Treaty* from Russia for the sum of $7,200,000.00 in gold. This is the only part of Arctic sovereignty acquired by purchase.

Soon after the Alaska Treaty, given the initial lack of effective administration of the US government in Alaska, rogue sealing ships from many nations engaged in extensive hunting and killing of fur seals, which had breeding grounds located in the Alaskan Archipelago. As a method of improving their administration over Alaska, Washington passed an 1870 legislative act addressing sealing, restricting catch numbers, hunting periods and granting a 20-year lease to the Alaska Commercial Company, a monopoly first ignored until the US Navy enforced the 1870 legislation and confiscated the equipment of all rogue sealers. As a significant number of these sealers were citizens of Canada, the British government responded to the pressures of the Canadian government to protest about US policy and the confiscations. However, the US maintained that because Russia had formerly claimed sovereignty over the entirety of the Bering Sea, the US had inherited the authority to exercise sovereign jurisdiction over their newly acquired half of the Bering Sea. Disagreement ensued and the US agreed to go to international arbitration over the issue.

The issues presented to the Paris Court of Arbitration in 1892 included that of whether Russia had maintained exclusive jurisdiction over the Bering Sea, and, as a result of the Alaska Treaty, whether 'the United States has any right, and if so, what right of protection or property in the fur seals ... when such seas are found outside the ordinary three-mile limit'.[66] The US never tried to prevent the freedom of navigation, nor the right of other nations to fish upon the high seas, but had only attempted to protect the fur seal from extinction. The arbitration court found the US in violation of the principles of the freedom of the seas and that the US had no right to confiscate the equipment of pelagic fur sealers and was required to compensate those whom they had economically disadvantaged.

So despite their apparent initial legal loss, the international legal challenge of US domestic legislation resulted in the creation of international environmental legislation on the basis of stewardship, largely considered to be the first case for stewardship of natural resources in international law.[67] The political and legislative events following the Alaska Treaty also led to the creation of the idea of environmental stewardship and the notion of the protection of the common heritage of mankind that would eventually become imbedded in the *United Nations Convention on the Law of the Sea 1982*. Of striking interest, Canada, the 'belligerent' against the stewardship claim of the US regarding the fur seals, now uses the same stewardship argument to promote their control of the Northwest Passage.

The year 1867 was also significant as it signalled the decline of British sovereignty over the region with the creation of the Dominion of Canada. From this point, Canada would incrementally acquire additional British territories, slowly changing the shape of future Arctic sovereignty, as the

Arctic was transferred from British to Canadian control in 1880 with the transfer of British Arctic islands. However, there was a significant issue in this transfer because not all 'Canadian Arctic' territories included in this transfer belonged to Great Britain by virtue of discovery and effective occupation. While much of this issue was settled through the exchange of diplomatic notes in which the US acknowledged that they did not wish to maintain claims on any discovery of islands in the Canadian Arctic Archipelago, this transfer of incomplete sovereignty would create a legacy of uncertainty for Canadian Arctic sovereignty in future Arctic affairs.

The development of the laws of territorial acquisition and events that occur within this period have several implications for the Arctic region, even though the political-legal aspects of many of the events themselves occurred outside of the physical Arctic. The first is that these laws and the processes of enforcing them established 'ownership' and subsequently the sovereignty of states over North American Arctic territories. Had these laws not been developed, indigenous populations would have retained the ownership of the land, but as European empires deemed that the land was not 'effectively occupied' for legal purposes, they were able to assume administration and ownership of the whole of the North American Arctic. The principles of territorial law continued to be reinforced through the granting of territorial charters, such as the transfer that gave all of Rupert's Land to the Hudson Bay Company, and also through treaties granting restoration of illegitimately seized land, as occurred in the *Treaty of Utrecht*. Finally, the laws of territorial acquisition provided the legal mechanism which determined the geopolitical boundaries that are seen today in Arctic politics and also created principles of utility and occupation that continue to be relevant in contemporary Arctic sovereignty debates.

Arctic sovereignty in the Interwar Years: 1908–45

The Interwar Years of the international system brought with them a change in the focus levied on the Arctic by the states that claimed and performed sovereignty over its territory. Although exploration continued in the Arctic region, sometimes sponsored by states, but often funded by private investors, the nature of the exploration transitioned from the pure pursuit of territory to scientific exploration; this scientific exploration was also used in pursuit of territory. Into the 1930s it was believed there might be additional yet undiscovered islands hiding within the sea ice and the new technologies of the period were used to search for these unclaimed territories. However, aside from the possibility of these few islands, all known land within the Arctic was claimed by the states of Canada, Russia, the United

States, Denmark and Norway. This period brought with it new conversation in territorial possibilities, such as the potential of sovereignty over sea ice, the sector principle and some of the final determinations of sovereignty over land caught up in the decline of the empire.

The acquisition of the North Pole in 1909 for the US heralded the end of the age of territorial discovery in the Arctic when Peary claimed the pole. However, the US rejection of the opportunity to claim sovereignty of the North Pole on the basis that it was located in the middle of the Arctic Ocean, on sea ice and part of the high seas established precedence within international relations that this aspect of the pole was outside the realms of territorial claims by sovereign states.

Meanwhile, aviation technology was being used by the Arctic states to determine if any more undiscovered islands existed within the Arctic Circle. This exploration activity, coupled with the tangled web of existing sovereignty over minor islands via discovery and the uncertain sovereignty due to inconsistent effective occupation of the minor Arctic islands, created a condition of insecurity over the northern territorial borders in the Arctic. To address this issue, first the Canadians issued a sector declaration in 1924 and began producing cartographies that delimit their territory to the North Pole, a practice that continues today. Then, in 1926 the Soviet Union issued a proclamation introducing the concept of the sector principle to the international system to address the circumpolar nature of the Arctic region. They said: 'All discovered lands and islands, as well as all those that may in the future be discovered ... are declared to be territories belonging to the USSR....'[68]

Equivalent to the notion of the 'regions of attraction' included in the US Monroe Doctrine, the sector principle indicated that all states were to relinquish claims to sovereignty over areas within this sector. Within the Canadian sector, the US were happy to comply as the only territory to which the US had any claim were minor islands within the Canadian Archipelago. This proclamation also had the result of solidifying Canadian claims to sovereignty over territories included in the 1880 Arctic islands transfer from Great Britain, who had not fully established sovereignty over all of the islands in the Canadian Archipelago through either discovery or effective occupation. Although the sector principle had the result of establishing regions of attraction within the Arctic Circle up to the North Pole, the sector principle failed to become the dominant rule for establishing territorial sovereignty over maritime areas within the Arctic region and only applied to terra firma within those sectors.

The final determinations of sovereignty over land within the Arctic occurred in the *Svalbard Treaty 1920* and the judgement on the *Legal Status of East Greenland 1933*. In both locations, multiple states had been utilising the land and the resources, eliminating Norway and Denmark from establishing clear sovereignty over the territory and thus it was necessary to settle the issue through legal mechanisms of the international system. The *Svalbard Treaty* provided for all member states already utilising the Svalbard islands to 'carry on there without impediment all maritime, industrial, mining and commercial operations on a footing of absolute equality'.[69] So, although the treaty gave 'most' sovereignty to Norway, it prevented it from having permanent sovereignty over the resources of Svalbard, usually understood to be a condition of territorial sovereignty.

In Greenland, due to historical divisions of empire in Scandinavia, such as the demise of the Kalmar Union, the legal status of this large Arctic island had failed to be established through the performance of effective occupation of a single sovereign state. The discussion over the sovereignty of Greenland was a matter of contention between Denmark and Norway, with Norway making declarations of sovereignty over the area of East Greenland. The International Court of Justice ruled in the judgment that the entirety of the island fell within Danish sovereignty, concluding the determination of the final major land borders of the Arctic region.[70]

There are several implications for Arctic sovereignty in the development of rules and codes during the Interwar Years. The first is that the Arctic states began to create innovative solutions for dealing with the circumpolar nature of the Arctic region. These include the use of the sector principle, although this method failed to establish sovereignty over maritime areas due to lack of universal application. A second implication is that it was also determined through political performances that sovereignty over the North Pole at the surface of the Arctic Ocean could not be claimed by states due to the pole's position in the middle of the high seas. The introduction of this rule of 'no sovereignty over sea ice' left open the other legal layers of the North Pole that would eventually be developed as the international system evolved throughout the twentieth century.

Arctic sovereignty in the Cold War: 1945–90

Sovereignty over the Arctic did not experience any changes during this period; however, the Cold War years brought with it radical change in the rules that frame the sovereignty of Arctic maritime spaces as the focus of the states turned to the resources of the world's oceans. The interstate conflict of the Great Wars during the Interwar Years had emphasised the need for states to secure new sources of energy, to maintain energy security for defence and for economic growth. In addition to the need for additional energy, the Interwar Years had seen vast developments in technology, creating more opportunity and options for the extraction of resources in previously unreachable places – especially within the oceans. These interests and capabilities of states introduced an opportunity for a major shift within the rules of the international system governing the maritime regions. The simple rules in the law of the sea, *mare liberum* and *mare clausum*, were about to undergo a major transformation.

This transformative shift in the rules organising the behaviour between states in the maritime was introduced by the US *Truman Proclamation 1945*, claiming that:

> United States regards the natural resources of the subsoil and sea bed of the continental shelf beneath the high seas but contiguous to the coasts of the United States as appertaining to the United States, subject to its jurisdiction and control.[71]

With this proclamation, the areas over which a state could claim sovereign jurisdiction expanded instantly to include large expanses of submarine territory. Like the introduction of all rules, this required international consensus to become part of the fabric of the international system, and as most other states in the greatly expanded international system saw the advantages of this rule to their own national interests, the rule quickly became a part of customary international law. This is why the Arctic states have turned to the new framework of the law of the sea to establish sovereignty over the maritime Arctic.

This rule of the continental shelf was soon codified in international law with UN Conventions adding new legal spaces of territory to the body of international law in territorial waters: the contiguous zone and the high seas. These new legal spaces make it possible for the Arctic states to claim relative sovereignty over maritime territory in the Arctic. However, as the Arctic states wait for their claims of up to 350 nautical miles to be reviewed by the United Nations Commission on the Continental Shelf, they continue to reiterate their historical borders established through the sector principle. Thus, Canada and Russia are now staking claims to multiple legal layers of space at the North Pole: through their claims to the Commission through the *United Nations Convention on the Law of the Sea 1982* and with their continuing diplomatic statements to sovereignty over the pole.

The final overarching rule of the international system during the Cold War Period was the rule of the existence of the Cold War itself. This rule existed not in codification, but only in the understanding of the behaviour of states interacting with one another in the international sphere – fuelled by physical insecurity and suspicions as each side promulgated their political ideologies to a wider audience. For the Arctic, which was the shortest route between the two sides of the Cold War conflict, this resulted in the region being seen as both an arena for warfare and also a direct route of attack. This caused many displays of military prowess and demonstration of capability to operate in the region. While the Cold War no longer exists in subjective understandings, the technology developed in this period equips the states to defend their prospective Arctic claims in the maritime regions.

The implications for Arctic sovereignty in the development of rules during the Cold War years include that the period was marked by several major transitions. The first includes the addition of new rules to the law of the sea, making it acceptable to appropriate resources of the maritime regions once considered as outside the jurisdiction of states. The development of the new layers of legal territory within the law of the sea was facilitated by the rise of liberal institutionalism following the Interwar Years, addressing not only the need to mediate global conflict, but also to address the growing number of states interacting within the sphere of the international system. Finally, although the sector has predominantly given way to the new rules of the law of the sea, these lines remain incorporated into performances of political technologies of sovereignty for the region.

Arctic sovereignty in the Contemporary Period: 1990–ongoing

The post-Cold War years of the international system are bringing with them changes in the way that the Arctic states engage with each other as the rule of the understanding of 'Cold War' has disappeared. States are increasingly cooperating on environmental matters for the region and have also developed a new framework for organising political action within the Arctic in governance through the Arctic Council. In this period, nearly all the bilateral boundaries (those not including the overlapping claims at the North Pole) are also being codified within international law. Thus, it seems that in this period, the codes and rules framing Arctic sovereignty continue to be developed within the evolving interests of states in their actions with one another and in their interaction with the larger global order.

A significant change for the Arctic is that of the establishment of the Arctic Council, emerging from the *Arctic Environmental Protection Strategy 1991* between all eight of the Arctic states. The cooperation provided through this agreement so near to the end of the Cold War gave states a platform for launching the Arctic Council, originally intended as an informal forum for discussing issues relating to the Arctic. However, in 2011, the Arctic Council was made into a formal organisation with the establishment of a permanent secretariat, with the additional intent of coordinating action between the Arctic states. The addition of this forum to Arctic political affairs adds a new layer of complexity, as it deals with issues outside of the jurisdiction of state sovereignty, providing a mechanism for states to project their Arctic specific interests into cooperation with other Arctic states. The governance of the Arctic Council allows for states to exert sovereignty over extra-territorial spaces.

The implications for the international system, and specifically for the development of the Arctic in this period includes the turn towards environmental cooperation in the region and the development of mechanisms of cooperation based on regional interests, developed outside of the institutions formed in the Cold War Period. This turn towards cooperation is the result of the diplomatic stand-down of Arctic militarisation, although the technology for this still exists. States still exercise their ability to operate in the Arctic in a military capacity, but some of this is in cooperation with one another as seen in the *Search and Rescue Agreement 2011*. This cooperation in the development of regulatory frameworks and military cooperation allows the Arctic states to form a coherent response to the interests of non-Arctic states in the resources and shipping lanes of the Arctic Ocean. Finally, the impetus to solidify maritime boundaries for resource security in the newly created territorial legal spaces in the Artic has been a continuing focus of the Arctic states.

Conclusions

So, who owns the Arctic?

It is easily apparent after reading this story of historical evolution of sovereignty that there is not a scramble for territory in the Arctic. The contemporary claims by the states littoral to the Arctic Ocean can be understood by tracing the development of political and legal events from the discovery of the New World to recent developments within international law. This is true both in claims to land and in their claims to the territorial division of the sea.

The Arctic has been owned by different state actors operating under different rule structures over the last several hundred years. The discovery of the New World by Europeans created a vacuum within the international system in which shared understandings of expected behaviour by states expanding their political boundaries into the territories of faraway lands were undefined. The creation of this vacuum acted as a catalyst for the introduction of a new set of rules to the international system. Although the unincorporated world was initially divided between Spain and Portugal by the accepted international authority of the period, within a century the legitimacy of this rule was challenged by other rising powers in Europe as the power of papal influence diminished in correlation with the rise of the notion of territorial sovereignty and the right of monarchs to be the ultimate authority within their territorial boundaries. Throughout this period, Great Britain and Russia dominated control over territories in much of the Arctic.

This change to the structure of the international system also brought with it a rising emphasis on the role of effective occupation and government administration in maintaining control over the territories of the Arctic. This transition in the normative understanding of the maintenance of sovereignty saw states using a new form of administration to demonstrate their authority over territory: the use of commercial trading companies to exercise economic engagement with the region. This was often used alongside more overt forms of effective occupation: legislative mechanisms, taxation, reorganisation of indigenous peoples and colonisation. Yet this was not the end of the evolution of effective occupation and the performances of sovereignty in the Arctic.

Within the rules-based and socially constructed international system, the shape of sovereignty in the Arctic would continue to respond to the introduction of new normative expectations. First, sovereignty over the Arctic would see a period where the extension of sovereignty beyond the edges of land and into the Arctic Ocean was tested. This sovereignty over the maritime was first denied when Peary's acquisition of the North Pole was not accepted as a recognised 'first discovery' because it was known that its location was within the frozen ocean. Arctic maritime sovereignty was again tested in the sectoral principle. Although both Canada and Russia still use these sector lines in contemporary cartographies, they have no standing in international law.

Sovereignty over the Arctic maritime found its legal footing in the evolution of the law of the sea. As this is a global standard for ocean governance, its universality naturally included the maritime spaces in the Arctic. It is this normative order that has made it possible for the Arctic states to claim new territory in the region and it is also the source of the rumoured 'scramble for territory'. So, while the Arctic states wait for their maritime territorial claims in the Arctic to be determined through the current rules of the international system, they continue to make diplomatic gestures using old familiar rules, although some of them have expired, to demonstrate the validity of their claims to the North Pole to the international community of states.

As the international system continues to evolve, so also change the rules and codes that guide the behaviour of states within this structure. Although the maritime territorial sovereignty of the Arctic is being determined through the current rules and codes of the international system, this is not the end of the evolution of Arctic as a region within the system. As new resources and new technologies are discovered, new layers of territory become accessible and new voids are identified in the legal structure of the international system. At present, the Arctic states are cooperating to make a new governance order for the region under the umbrella of the Arctic Council. Through this forum, they are developing governance for the Arctic and participation in this organisation requires non-Arctic states to recognise the sovereignty of the Arctic states over the region.

This will not be the end of the story of who owns the Arctic. The ongoing development of Arctic specific rules and codes will continue to be the product of the needs of the Arctic states in response to the constraints of the international system in which the guidelines for sovereignty are developed. Watch this space.

Notes

1 Lawson W. Brigham, 'Think Again: The Arctic', *Foreign Policy*, October 2010, p. 71.
2 Chet Van Duzer, 'The Mythic Geography of the Northern Polar Regions: Inventio Fortunata and Buddhist Cosmology', *Culturas Populares* 2 (2006): 1–16. pp. 2–3.
3 Clements R. Markham, 'Pytheas, the Discoverer of Britain', *The Geographical Journal* 1, no. 6 (1893): 504–24.
4 H.F. Tozer, *A History of Ancient Geography*, Cambridge Geographical Series (Cambridge: Cambridge University Press, 1897), p. 160.
5 Robert McGhee, *The Arctic Voyages of Martin Frobisher: An Elizabethan Adventure* (Montreal, Quebec: McGill-Queen's University Press, 2001).
6 Inga-Maria Mulk and Tim Bayliss-Smith, 'Liminality, Rock Art and the Sami Sacred Landscape,' *Journal of Northern Studies* 1, no. 2 (2007): 95–122, p. 107.
7 Samuel Eliot Morison, *The Great Explorers: The European Discovery of America* (New York: Oxford University Press, 1978).
8 Stephen Van Evera, *Guide to Methods for Students of Political Science* (Ithaca, NY: Cornell University Press, 1997), p. 70.
9 The US federal government owns 62 per cent of land in Alaska.
10 For example, see the Nunavut Land Claims Agreement 1993.
11 Balibar says 'collective property and particularly state property ... is in itself nothing but private property....' Etienne Balibar, *Studies on Politics and Philosophy Before and After Marx*, trans. James Swenson (New York: Routledge, 1994), p. 218.
12 'Sovereignty', *Oxford Dictionaries*, 20 August 2017, https://en.oxforddictionaries .com/definition/sovereignty.
13 Robert H. Jackson, *Sovereignty: The Evolution of an Idea* (Cambridge: Polity Press, 2007), p. ix.
14 See Stephen D. Krasner, *Sovereignty: Organized Hypocrisy* (Princeton, NJ: Princeton University Press, 1999).

15 See Wendy Brown, *Walled States, Waning Sovereignty* (Brooklyn, NY: Zone Books, 2010).

16 Stephen Krasner has identified four types of sovereignty: international legal sovereignty, Westphalian sovereignty, domestic sovereignty and interdependence sovereignty. See Stephen D. Krasner, *Sovereignty: Organized Hypocrisy*, p. 3.

17 Niccolò Machiavelli, *The Prince*, trans. J.S. Byerley (London: Sherwood, Neely & Jones, 1810), p. 23.

18 Francesco Maiolo, *Medieval Sovereignty: Marsilius of Padua and Bartolus of Saxoferrato* (Delft, The Netherlands: Eburon Academic Publishers, 2007), p. 109.

19 Jean Bodin, *On Sovereignty* (Cambridge: Cambridge University Press, 1992), p. 1.

20 Ken Macmillan, *Sovereignty and Possession in the English New World: The Legal Foundations of Empire, 1576–1640* (Cambridge: Cambridge University Press, 2006).

21 David Armitage, 'John Locke: Theorist of Empire?', in *Empire and Modern Political Thought*, ed. Sankar Muthu (Cambridge: Cambridge University Press, 2012), 84–111, p. 104.

22 Kent McNeil, 'Sovereignty and Indigenous Peoples in North America', *Articles & Book Chapters* 2642 (2016): 82–103, p. 82.

23 Stephen D. Krasner, 'Sovereignty', in *Blackwell Encyclopedia of Sociology*, ed. George Ritzer (Oxford: Blackwell Publishing, 2007), p. 4632.

24 See, for example Pärtel Piirimäe, 'The Westphalian Myth and the Idea of External Sovereignty', in *Sovereignty in Fragments: The Past, Present and Future of a Contested Concept*, ed. Skinner Kalmo (Cambridge: Cambridge University Press, 2010), 64–80.

25 Pärtel Piirimäe, p. 68.

26 Morris R. Cohen, 'Property and Sovereignty', *Cornell Law Quarterly* 13 (1927): 8–30, p. 9.

27 See Jeffrey Glover, *Paper Sovereignty: Anglo-Native Treaties and the Law of Nations, 1604–1664* (Philadelphia: University of Pennsylvania, 2014).

28 Such as Venice and Genoa. See Bo Johnson Theutenberg, 'Mare Clausum et Mare Liberum', *Arctic* 37, no. 4 (1984): 481–92.

29 Jeffrey Glover, *Paper Sovereignty: Anglo-Native Treaties and the Law of Nations, 1604–1664*, p. 160.

30 Hugo Grotius, *The Freedom of the Seas* (New York: Oxford University Press, 1916), p. 69.

31 Anthony Scott, *The Evolution of Resource Property Rights* (Oxford: Oxford University Press, 2008), p. 136.

32 Friedrich August Freiherr von der Heydte, 'Discovery, Symbolic Annexation and Virtual Effectiveness in International Law', *The American Journal of International Law* 29, no. 3 (1935): 448–71, p. 470.

33 W. Lakhtine, 'Rights Over the Arctic', *The American Journal of International Law* 24, no. 4 (1930): 703–17.

34 Kent Burch, 'Toward A Constructivist Comparative Politics', in *Constructivism and Comparative Politics*, ed. Richard T. Green (New York: M.E. Sharpe, Inc, 2002), 60–87, p. 75.

35 Kurt Vonnegut, *Breakfast of Champions* (New York: Random House, 1973), p. 10.

36 Friedrich August Freiherr von der Heydte, 'Discovery, Symbolic Annexation and Virtual Effectiveness in International Law', p. 454.
37 Arthur S. Keller, Oliver J. Lissitzyn, and Frederick J. Mann, *Creation of Sovereignty Through Symbolic Acts 1400–1800* (New York: Columbia University Press, 1938), p. 144.
38 Friedrich August Freiherr von der Heydte, 'Discovery, Symbolic Annexation and Virtual Effectiveness in International Law', p. 454.
39 Andrew F. Burghardt, 'The Bases of Territorial Claims', *Geographical Review* 62, no. 2 (1973): 225–45, p. 229.
40 Andrew F. Burghardt, p. 228.
41 W. Lakhtine, 'Rights Over the Arctic', p. 705.
42 Columbus had local knowledge of Portuguese possession off the Africa coast as he once lived on the island of Porto Santo.
43 Arthur Davies, 'Columbus Divides the World', *The Geographical Journal* 133 (1967): 337–44, p. 339.
44 Perhaps 'rediscovered' is a better term as when Columbus arrived in the New World, indigenous populations were already well established. There is also evidence that Benedictine Irish Monks and the Vikings all found North America before Columbus.
45 William Hickling Prescott, *History of the Reign of Ferdinand and Isabella the Catholic, of Spain* (London: Richard Bentley, 1838), p. 259.
46 Columbus is given credit for being the father of the Treaty of Tordesillas in Davies, 'Columbus Divides the World'.
47 Robert Phillimore, *Commentaries Upon International Law* (Philadelphia: T. & J.W. Johnston, 1854), p. viii.
48 Davies, 'Columbus Divides the World', p. 339.
49 Davies, p. 338.
50 Jesse S. Reeves, 'International Boundaries', *The American Journal of International Law* 38, no. 4 (1944): 533–45, p. 541.
51 Francis Gardiner Davenport, ed., *European Treaties Bearing on the History of the United States and Its Dependencies to 1648* (Washington DC: The Carnegie Institution of Washington, 1917), p. 95.
52 Wilhelm Grewe, *The Epochs of International Law*, trans. Michael Byers (New York: Walter De Gruyter, 2000), p. 157.
53 Davenport, *European Treaties Bearing on the History of the United States and Its Dependencies to 1648*, p. 260.
54 Ram Prakesh Anand, *Origin and Development of the Law of the Sea: History of International Relations Revisited* (The Hague: Martinus Nijhoff Publishers, 1982), p. 442.
55 Alan Cafruny, 'Class, State and World Systems: The Transformation of International Maritime Relations', *Review of International Political Economy* 2, no. 2 (1995): 285–314, p. 290.
56 Cafruny, p. 292.
57 Scott, *The Evolution of Resource Property Rights*, p. 136.
58 Lassa Oppenheim, *International Law: A Treatise*, 2nd edn (London: Longman, Green & Co., 1912), p. 319.
59 Elliot J. Feldman and Michael A. Golberg, *Land Rites and Wrongs: The Management, Regulation and Use of Land in Canada and the United States* (Cambridge, MA: Lincoln Institute of Land Policy, 1987), p. 37.
60 Charles Sumner, *Speech on the Cession of Russian America to the United States* (Washington DC: Congressional Globe, 1867), p. 4.

61 Sumner, p. 4.
62 Corine Wood-Donnelly, 'Exploring and Mapping Alaska: The Russian American Era, 1741–1867', *Terrae Incognitae* 49, no. 1 (2017): 99–101, p. 100.
63 John Bassett Moore, *History and Digest of the International Arbitrations to Which the United States Has Been a Party* (Washington DC: Government Printing Office, 1898), p. 755.
64 Henry W. Clark, *The History of Alaska* (New York: The Macmillan Company, 1930), p. 64.
65 James Alton James, *The First Scientific Exploration of Russian America and the Purchase of Alaska* (Chicago: Northwestern University, 1942), p. 20.
66 James Thomas Gay, *American Fur Seal Diplomacy: The Alaskan Fur Seal Controversy* (New York: Peter Lang, 1987), p. 73.
67 Malgosia Fitzmaurice, *Contemporary Issues in International Environmental Law* (Cheltenham, UK: Edward Elgar, 2009).
68 Jane Tabrisky Degras, ed., *Soviet Documents on Foreign Policy: 1925–1932* (London: Oxford University Press, 1952), p. 104.
69 The Svalbard Treaty (9 February 1920).
70 There is currently a dispute between Canada and Denmark over the sovereignty of Hans Island, a mere rock, located in the Davis Strait.
71 Harry S. Truman, 'Proclamation 2667 – Policy of the United States with Respect to the Natural Resources of the Subsoil and Sea Bed of the Continental Shelf', Document Archive, The American Presidency Project, 28 September 1945, www.presidency.ucsb.edu/ws/?pid=12332.

Part II

Policy and visual representations

Given the Arctic is an ice-covered ethereal space where the maritime sometimes functions as land, the state performance of sovereignty confounds the usual division of land and sea. Until Peary reached the North Pole, the exploration of the Arctic for both political and scientific purposes was never divided between land and sea in any sense of legal division and the ongoing mixed use of the region has perpetuated that sense of the Arctic as an integral political space. In its own unique way, the Arctic has defied normative classifications and state performances of sovereignty in policy practices are reflective of this condition. In the case of the Arctic, what should be domestic policies towards the region have a 'split personality' because they apply both to land, over which states exercise absolute sovereignty, and also to maritime spaces, over which states exercise only relative sovereignty and therefore the policies broach into the realm of foreign policy.

Although it was understood that the North Pole was located in the high seas well into the twentieth century, it was believed that underneath all that ice there could be another Arctic continent. So even as the vestiges of imperial methods for acquiring new territory by discovery were fading away as the entire globe became coloured into political spaces of empire, the rapid acceleration of aviation technology put the process of exploration in the Arctic on fast forward. Instead of the slow laborious trekking for days and weeks over snow and ice, or ships getting caught fast in sea ice, a plane could quickly fly from A to B through the Arctic, spotting potential unclaimed territories. It was this urgency that, in part, caused the proclamation of sectoral claims by Canada and Russia.

The point of all this explanation is to remember why the discourse on the geopolitics of the Arctic is often so confusing. It is so complex because the normative rules for territorial acquisition and effective occupation changed. As the means for exercising these rules advanced technologically, so also the methods that states have used to demonstrate the performance of their sovereignty and effective occupation over the Arctic evolved. Therefore the performances of sovereignty have become a complex set of arrangements of the signalling usage of both old and new rules with the political technologies of both old and new methods. The interesting part is that the Arctic states put considerable energy into demonstrating their sovereignty over the land and also their ability to exercise effective occupation over the whole of the Arctic. As such, there are still performances that can be interpreted as ceremonies of possession over new normative constructions of legal space, spaces which did not legally exist a century ago.

The performance of sovereignty in the Arctic almost always radiates from landed territories all the way to the symbolic point at the North Pole. Why is this? It is in part because old habits are hard to break. The nuanced nature of the ice-covered Arctic region along with the unknown extent of land made it difficult to know where to draw the limits for expressing and performing sovereignty. By the time that the physical geographic realities of the Arctic were known, new norms for maritime territory were being created in the law of the sea, with new extensions of sovereignty in spaces such as the continental shelf. While continental shelf claims are not supposed to be subject to the rule of effective occupation, there is an element of historical use that comes into play in the case of overlapping claims. These are the conditions that frame the geopolitical performances throughout Arctic space.

The Arctic is a space where policy and its application continues to respond to normative evolutions. Interestingly, we can see the manifestations of this policy in state performances of sovereignty happening across the circumpolar Arctic. How is this possible? Because these performances of sovereignty have been recorded in visual form by the tiniest messenger of the state: the postage stamp. As the smallest of all government documents, the postage stamps of a country often transmit messages about the interests, identities and policies of that state.

Most people will never read the international treaty on the protection of the polar bear, but they may see the polar bear on a postage stamp and learn that this iconic animal has an important protected status. Even the distinguished historian Hobsbawm indicated that he 'subconsciously absorbed the tangled political history of post-1914 Europe'; as a child collecting postage stamps his view on history was informed through the 'unchanging continuity of George V's head on British stamps and the chaos of overprints, new names and new currencies elsewhere'.[1] If he can absorb the history of Interwar Europe through the changing images on the postage stamp, then most certainly we can learn also about the interests of states in the Arctic in the same way. We only have to look for it.

The messages transmitted in this commercialisation of sovereignty have also come to the interest of researchers because beyond popular culture, stamps 'reflect ideologies, aspirations and values, attesting to political, social and cultural ideas and aesthetic tastes. When closely inspected, stamps often prove to convey much more than meets the eye'[2]; the 'meaning [can be] derived from the nature of the whole image and circumstances of its creation'.[3] So how can we connect the symbols and visual images on postage stamps to the known policy and performances of sovereignty in the Arctic?

The images on postage stamps offer themselves to the creation of themes and they are the reason that some countries produce stamps with appeal to thematic collectors. Described as the 'commercialization of state sovereignty', it is the reason that Mickey Mouse appears on the postage stamps of Bhutan, Elvis on the stamps of Burkina Faso and the French Antarctic features on a stamp from Monaco.[4] But in a more serious vein, states often use postage stamps as a way to signal participation in serious areas: civil rights, space programmes or international treaties. As a result, the creation of themes in postage stamps lends itself to a useful methodological and analytical tool that researchers refer to as data coding. Coding is a methodology that allows the research to gather data into groups because shared characteristics show an emerging pattern.[5] From these patterns the researcher can link data to patterns of action and interpret the codes related to empirical or theoretical knowledge.

Very useful to the study of Arctic policy, some of these messages are versatile and are subliminally influenced by the use of national symbols or other graphic references to national history, culture and geography. This use of symbolism to establish legitimacy is not uncommon in imperial practices as states have 'an acute awareness of the power of symbols to shape popular perceptions and to fuel national sentiment, and hence to shape the political reality'.[6] By presenting the visuals on the postage stamp as the symbols of the nation, the state is narrating the story of itself in pictorial form, and it is only left for the viewer to absorb and tacitly accept the message, legitimising the position of the state.

However, this research does not investigate the range of themes tempting collectors, but instead focuses on the theme of Arctic stamps produced by the US, Canada and Russia in order to evaluate them for evidence of effective occupation within their portrayal of the Arctic and to evaluate these themes for parallels to Arctic policy. The postage stamps did not disappoint and visual representations of discovery, annexation and the political technologies of effective occupation were present in the stamps depicting the Arctic. Following an examination of the entire collection of the postage stamps of these three states, it is clear that there have been transitions in the way the effective occupation is represented, and the trends closely correspond with changes in the rules and norms of sovereignty and territory in the international system, as well as the overarching political issues of the various periods. This includes issues such as representation of military prowess in the Cold War, but also changes in the attitudes toward indigenous people as they ceased to be part of the 'white man's burden' and instead are recognised as sovereign people with rights over their homeland territory, its resources and their self-determination.

By analysing and describing who or what is pictured on the Arctic stamps in their visual representations, five themes have been coded relating to the performances of sovereignty in the Arctic: (1) Nature, (2) Exploration, (3) Maps, Flags and Territory, (4) Effective Occupation and (5) Indigenous Inhabitants and Culture. These themes were selected to discern whether the methods of the performances of sovereignty, including exploration, effective occupation and asymmetric perceptions of the indigenous inhabitants 'provide for critical analysis of the social and power relations framing philatelic iconography'.[7] The results reveal much about the underlying strategies of the Arctic states in their policy towards the region, helping to visualise their methods for performing and maintaining sovereignty, because

> in over a century of commemorating the Arctic in postage stamps, categories relating to its exploration, occupancy, its people and nature, these states have illustrated a narrative of their relationship with the Arctic to both a domestic and to an international audience.[8]

The next three chapters marry the evolution of Arctic policy with the performances of sovereignty as seen in the visual representations of postage stamps of the US, Canada and Russia.

Notes

1 Eric Hobsbawm, *Interesting Times: A Twentieth Century Life* (New York: Pantheon Books, 2002), p. 314.
2 Ami Ayalon, 'The Hashemites, T.E. Lawrence and the Postage Stamps of the Hijaz', in *The Hashemites in the Modern Arab World: Essays in Honour of the Late Professor Uriel Dann*, eds. Asher Susser and Aryeh Shmuelevitz (Abingdon, UK: Frank Cass & Co, Ltd, 1995), p. 16.
3 Roland Barthes, *The Responsibility of Forms*, trans. Richard Howard (Oxford: Blackwell Publishing, 1985), p. 7.
4 Joel Slemrod, 'Why Is Elvis on Burkina Faso Postage Stamps?: Cross-Country Evidence on the Commercialization of State Sovereignty', *Journal of Empirical Legal Studies* 5, no. 4 (2008): 683–712, p. 684.
5 Johnny Saldaña, *The Coding Manual for Qualitative Researchers*, 2nd edn (London: Sage Publications, 2013).

6 Yair Wallach, 'Creating a Country Through Currency and Stamps: State Symbols and Nation-Building in British-Ruled Palestine', *Nations and Nationalism* 17, no. 1 (2011): 129–47, p. 138.
7 Daniel Hammet, 'Envisaging the Nation: The Philatelic Iconography of Transforming South African National Narratives', *Geopolitics* 17, no. 3 (2012): 526–52, p. 528.
8 Corine Wood-Donnelly, 'Messages on Arctic Policy: Effective Occupation in the Postage Stamps of the United States, Canada and Russia', *Geographical Review* 107, no. 1 (2017): 236–57, https://doi.org/DOI: 10.1111/j.1931-0846.2016.12198.x, p. 254.

2   The American Arctic

Introduction

As a political geographic space, the American Arctic in Alaska has two histories. It has an early history as part of the Russian empire and a second history as a territory of the US. First, the American Arctic was discovered, explored and mapped by Imperial Russia and its representatives. Trace evidence of this uncelebrated legacy remains, although this history is largely buried in Russian exploration archives, in part because of the 'inherent secrecy of Russian society'.[1] Mostly, however, little is communicated about this early heritage in US government communications because celebrating this legacy does not narrate the position of US sovereignty over Alaska. It is only the American performances of sovereignty, occurring as a result of policy agendas, that are celebrated in visual representations of US effective occupation. These performances can be seen in the US postage stamps depicting the semiotics of sovereignty in the Arctic.

From these images emerge a two-pronged narrative of US engagement with the region. The first is a narrative of how the US have adapted their projected communications of their sovereignty over Alaska in line with the transitions in the international system – from the changes in territorial expansion when the US acquired Alaska in a sovereign real estate deal to contemporary approaches in projecting their role as nature conservationists in a period when there is much tension over the development of fossil fuel resources in the Arctic. The second is a narrative of the political technologies that are significant policy instruments for securing that sovereignty. These policy approaches include environmental stewardship, security, territorial administration, historical and scientific exploration, and, infrequently, indigenous culture.

Arctic engagement in the twenty-first century shows the US to be engaging with the Arctic in an anchored, but arguably low-profile manner compared with the often brash activities and policy statements produced by Canada and Russia. This relationship with the Arctic is in part a result of the US internal political scene, where partisan and bureaucratic politics have hampered decisions on a range of issues from drilling for fossil fuels in the Arctic National Wildlife Refuge (ANWR) to the ratification of the *United Nations Convention on the Law of the Sea 1982* (UNCLOS). As a way of framing their sphere of interest, the US have one of the smallest of the Arctic sectors although this region also extends south of the Arctic Circle into the Bering Sea. In its maritime territory, Alaska has an unfortunate physical geography with a short continental shelf that likely will not result in extended territorial claims to submarine exploitation rights, even if the US were to ratify UNCLOS.

Holding a certain sovereign title over the territory of Alaska, US strategy in Arctic international politics pursues a different set of objectives to those of the other Arctic states, especially Canada and Russia, who are vying to demonstrate historical claims to the North Pole. While Canada and Russia pursue Arctic foreign policies to establish and extend sovereign claims in the Arctic, the US are pursuing an Arctic policy that fulfils broader US national interests including interest of security, the environment, and economic and infrastructural development. The modern policy is largely reflective of overall US Arctic policy approaches as it has deviated little throughout the periods of US engagement in the region. However, some of the methods for pursuing this policy have changed with the transitions within the international system and the formality of the relationship has changed from a status merely as an incorporated territory, to that of a full-fledged member of the union of the US.

US Arctic policy in the Age of Discovery

US engagement with the Arctic began during the latter end of the period of imperial expansion within the international system promoted by powerful states. This was spurred by two major concepts within US policy in the middle of the nineteenth century: Manifest Destiny and the Monroe Doctrine. Early thought within the US had determined that the Mississippi River offered a defensible and formidable natural geographic boundary, but with the almost accidental Louisiana Purchase, the territorial size of the US doubled overnight. Following this rapid territorial expansion by the US government, American settlement and economic activities flowed rapidly westward. They soon realised that European empires with territorial claims throughout North America were retarding US expansion 'limiting our greatness and checking the fulfilment of our manifest destiny to overspread the continent'.[2] When this expansionist policy was combined with the Monroe Doctrine, formally warning European powers that 'the American continents ... are henceforth not to be considered as subjects for future colonization by any European powers',[3] it was noted that the US believed the Pacific Ocean to be the new western limit of their boundaries. In this period, three key objectives emerge: civil and legal organisation, development of economic resources and conservation of resources for national security.

During this period William H. Seward, the US Secretary of State, was an advocate of US expansionist policies and pursuing territorial and resource acquisition. Seward attempted a failed purchase of Greenland, but successfully added the Alaska Territory to the US territorial holdings in 1867, for the sum of 7.2 million dollars in gold, or $0.02 per acre. This transaction ultimately signalled to other powers that might challenge US territorial expansion that the US government was sufficiently solvent, even following the costly domestic civil war, to make extravagant purchases in gold. This purchase implemented the Monroe Doctrine by removing a European power completely from the landscape of North America. Because of his contributions to the expansion of US territory, Seward is commemorated on the first US Arctic postage stamp in 1909.

Ratification for funds to conclude the *Alaskan Purchase Treaty* was delayed for month by Congress being divided over whether the empty 'wasteland' of Alaska was worth the price tag set by the Russians when there was a chance the land could simply be abandoned, and therefore free, in the near future. As the entire Alaskan region was believed to be icebound and desolate, political opponents labelled the purchase 'Seward's Folly'. However, Seward believed in American expansion and saw political and economic value in the new territory, which had scarcely been utilised by Imperial Russia due to corruption by the Russian-American company. Visiting the new territory in 1869, Seward said, 'I realized, if indeed I did not discover in those [Alaskan] Territories a new, peculiar, and magnificent field of commerce and empire.'[4] And so, the vision of an American imperialist, pursuing the national interest of territorial expansion, made the US into an Arctic state long before the true security and economic values of the region were realised. Despite the apparent folly of the purchase, manifest destiny and the desire to maintain positive diplomatic relations with Russia caused Congress to approve the investment capital for Alaska.

Within the space of a few years, rather than the generation of time that Seward predicted, the economic value of the Alaskan Territory was realised. It was written of this early maturation of Alaskan economic value that:

our new possession has seemed almost a myth too vague and intangible ... like an unexpected legacy, its magnificence and value have not yet been comprehended; but the time is close at hand when her mighty forests will yield their treasures, her mines will open out their richness, her seas will give of their abundance, and all her quiet coves will be converted into busy harbours.[5]

Within two decades, the purchase cost of the territory was regained from the trade of seal furs alone,[6] and by the 1950s, Alaska had paid for itself 425 times over![7] It was through the economic development of Alaska and the charter of the Alaskan Commercial Company in 1870 that the US first demonstrated sovereignty over Alaska in the same method as Imperial Russia had done through the Russian-American Company.

The *Alaskan Treaty Purchase*, as the first US policy document for the Arctic territory, outlines the broader US interests of the period for the region. In practice, the treaty defines the boundaries of the territory, gives the terms of the cession in property and commercial contracts, and details the treatment of foreign citizens, natives and military installations. With its ratification, the US federal government immediately assumed authority and sovereignty over all *terra firma* in Alaska. However, the US extended 'the laws of the United States relating to customs, commerce, and navigation';[8] it initially faced difficulty in the development of effective occupation because of the lack of colonists in the territory. It also continued the paternalistic and asymmetric relationship the US held with other native tribes in the continental US, establishing a custodial relationship between the US and the indigenous inhabitants of Alaska.

Due to the small population, Alaska was first organised as a military district, although the desire to develop the economic resources of the region would create a sense of urgency in expanding government administration throughout the region. For several decades, Congressional and Presidential communications would frequently refer to the lack of civil administration in Alaska. While the US Constitution enables the federal government to extend laws and regulation to a new territory, this system was initially problematic in Alaska when

the total white population of Alaska is about 250, and for purposes of political illustration, the number of voters is usually put down at 15. To find this handful of people a Governor and a representation in Congress, to say nothing of the courts, would be a farce of the broadest kind.[9]

Congress strengthened US sovereignty over the region when it established civil and legal governance institutions in the *1884 Alaska Organic Act*.

After the population rapidly grew during a regional gold rush, the territory was formally incorporated as US territory in the *Second Organic Act for Alaska* in 1912. With this, government administration over Alaska was complete.

The development of US economic interests in the Arctic with the establishment of the Alaska Commercial Company led to legislative protectionism over commercial resources with tensions arising between the development of private interests and public interests in both terrestrial and maritime areas. In landed spaces, the expansion of private interests saw wealthy entrepreneurs creating railroad and commercial monopolies by buying up large swathes of resource rich land. Yet the creation of Alaska as an incorporated territory where 'conservation won, the corporation was halted'[10] significantly needed 'the development of agriculture [to] proceed with the opening of and development of the country'.[11] In this period several nature reserves were established in Alaska and vast swathes of forested land were set aside for future generations. Under the umbrella of conservation, even if only rationalised for the protection of strategic economic resources, environmental stewardship became a feature of US policy in the Arctic.

This policy feature became a point of friction between commercial and public interests when tensions with Canada emerged over the US attempt to restrict fur sealing in the Bering Sea beyond the limits of the customary three nautical miles of territorial waters. By claiming sovereign authority over the Bering Sea, the US attempted to protect the commercial longevity and continuation of fur seals as an economic resource through domestic legislation and enforced this in extended maritime spaces. The issue was resolved by an international tribunal who determined 'the United States had no rights of protection or property'.[12] These developments reinforced the principle of the freedom of the seas, but promoted both domestic and international legislation based on environmental consideration, a trend that continues in performances of sovereignty across the Arctic.

Despite the neat transfer of territory from Russia to the US, some existing territorial rivalries remained unresolved. For example, the territorial land boundary between Imperial Russia and Canadian British Columbia had never been clearly delimited due to disagreement over the interpretation of an 1825 boundary treaty, which remained unresolved in the 1867 Alaskan Purchase. Accelerated by realisation of the economic potential, the Alaskan boundary dispute between Canada and the US emerged. In 1903, the position of the boundary was agreed in a politically contested arbitration deal involving Canada, the US and Great Britain, with the British siding with the US due to other global political issues. However, the contemporary Beaufort Sea dispute is a continuation of the same disagreement over the interpretation of the 1825 treaty, and the issue remains for the US to resolve their finalised Arctic territorial claims and resource rights.

Throughout these years of territorial development in Alaska, the US continued to finance expeditions into the High Arctic, including expeditions to the North Pole. The motivations for the expeditions included access to commercial passages, procurement of resources and, ultimately, the motive of discovery: possession.[13] During this period of expansion, the US acquired 'first stage' possession of many islands in the Canadian archipelago. However, by the time Peary reached the North Pole in 1909, the tide of territorial expansion through discovery was turning in the international system, affecting the spread of sovereignty claims in the Arctic. Although Peary claimed the North Pole for the US, expensive lessons from declaring jurisdiction over fur seals in the Bering Sea had just been learned. So, the US government declined the opportunity to accept this possession.

US Arctic policy during this period of discovery saw the development of several objectives for the US government including the expansion of territory, extension of the functions and mechanisms of government over the new territory, establishing control over resources and the exploitation of their economic potential. Additionally, the government reinforced the asymmetric relationship with the indigenous populations maintained elsewhere in the US by excluding them from the rights and privileges of ordinary US citizens, instead absorbing them within an unequal custodial relationship. The progressive finale of this period also saw the development of environmental stewardship agendas, even if only for the protection of commercial interests. This practice triggered the beginning of environmental conservation and preservation principles moving into the realm of international politics.

The commemoration of Seward in this early postage stamp reflects the US relationship with Alaska in a fitting, although understated, memorial to the expansionist policies of the US in respect of their national interests in the Arctic. With the last of all known lands around the globe having been discovered, recorded and mapped within the political jurisdiction of states worldwide, the Age of Discovery expired. The US entered the stage of the emerging geopolitics of the Arctic, facilitating the preservation of the notion of the freedom of the sea both in their declination of sovereignty over the North Pole, but also in their defeated challenge to sovereignty over the Bering Sea. The solidification of these norms of sovereignty for maritime spaces would propel the policy agenda of the US parallel to the exigencies of the next phase of the international system overlapping with the performances of sovereignty in the Arctic: The Interwar Years.

US Arctic policy in the Interwar Years

The US issued their next Arctic stamp in 1937, the only image produced in the Interwar period, commemorating the creation of Alaska as a US territory. It is an image that presents an idyllic portrayal of the perfect sovereignty of the US over Alaska: it is the dream of the American frontier, settled and effectively occupied, within the frame of a postage stamp. The image shows the lush and excellent conditions of agriculture as a farmer works his well cultivated fields. He stands with a cosy farmhouse against the backdrop of Alaska's natural beauty with snow-capped peaks and an expansive lake. Coming in the middle of the Interwar period, the idea this image portrays is a fiction at some levels; it hides the situation that government policy planners were overcoming with their Arctic policy strategies. The policy priorities that dominate this era focus on domestic development of Alaska and broader security issues of the US.

US policy and engagement with the Arctic transitioned during the Interwar Years, moving to the use of the American Arctic as a tactical cache of resources, but also using the entirety of the North American Arctic as a security buffer zone and staging point for military operations. Exploration and the impulse for discovery continued, but as technology such as aeroplanes and radio communications were developed, these tools were used to pursue strategic US interests, in science, economics and security. The development of Alaska and the transitions in the effective occupation of the territory was a slow undertaking by the US government, despite their active hand in policy and planning with the intent of utilising the resources and benefits of the region. When Alaska fully incorporated as a territory of the US, the early method of using state-chartered commercial companies to exercise effective occupation turned towards other methods: private resource development and settler colonialism. As land was made available for purchase from the government by private investors, settlers and major industries such as fishing, canneries mining and large-scale reindeer farming contributed to the economics of the continental US from Alaska.

While these economic activities were beneficial for their rich investors, they were less effective methods of settling an expansive territory and Alaska struggled to reach adequate population levels for economic and territorial occupation. The first problem is that many of these industries attracted only male workers; the ratio of men to women in the territory was always high – from over 2.5:1 around the turn of the twentieth century to an average of 1.25:1 throughout the Interwar period.[14] This highly male and highly transient population did not lend itself to creating stable communities throughout Alaska and the various boom-and-bust cycles of these industries did little to attract those who would ordinarily provide community services. As government land-use policy shifted towards conservation and when the Great Depression caused global markets to fail, many of the resource extraction industries collapsed and the workers left Alaska.

Despite the wide scale availability of arable land through government policy, few adventurous farmers decided to move to Alaska, in part because of the predominating myth that Americans 'want to believe that Alaska is a land of snow and ice'.[15] US census records show a staggeringly low number of farms in Alaska in the period, contrary to analysts' predictions. In the newest and richest of US frontier landscapes: 'the almost complete failure to attract settlers seems attributable mainly to a decline in the traditional advantages of frontier agriculture in the twentieth century and a simultaneous rise in the opportunities and amenities of urban life'.[16]

By the Interwar period, urban areas with their vast modern opportunities and lifestyles were making farming and the homestead an increasingly old-fashioned choice. US policy under the New Deal Roosevelt Administration established 'government-assisted colonists in the most suitable region of Alaska' because this 'settlement would aid the growth of the territory, supply agricultural foodstuffs for Alaska, and serve as a step toward the possible defenses of the region'.[17] Yet even these government sponsored programmes to improve the settlement of Alaska nearly failed until the issues of national security made them economically profitable as they supported the influx of military installations into Alaska.

Another significant aspect of US policy towards the Arctic was the focus on resource development and the installation of military infrastructure to promote national security capacities. In tangent with these objectives, Arctic exploration did not disappear completely, but the aims of exploration turned from 'discovery' of territory towards exploration for scientific and economic aims because 'Alaska, as a storehouse, should be unlocked'.[18] Funding from the US government enabled scientific exploration due to 'direct bearing upon economic values sooner or later'.[19] This scientific exploration of the Arctic enabled the pursuit of both the economic and security interests of the US in the region with tax funded agencies such as the US Geological Survey, the Alaska Mining Commission and the Air Corps of the US Army. These administrative departments began mapping the Arctic terrain on both land and sea, locating natural resources and installing navigational aids.

The policy of funding both infrastructural development and scientific research in the Arctic, now considered standard practice by the US government, aided in the development of military infrastructure, including roads, radio communications and airstrips. The US also participated in the Second International Polar Year in 1932, collaborating on research to solve the abnormalities of magnetic field lines in the Arctic environment that caused interference in radio communications. Congress approved a project for 'furnish[ing] a route to link up the established airfields and thus permit their expansion' and to 'provide an overland auxiliary supply route to Alaska'[20] with a transcontinental road across Canada from the continental US to Alaska. Following the war, the US surrendered control of the Canadian segment of the highway.

Despite the embarrassing occupation of the Aleutian Islands by Japan during the Second World War, US policy also supported operational and training research for the troops. Commissioned by the War Department, polar explorer Stefansson produced a military Arctic manual because 'there might be a war coming which would carry our troops to spheres of operations where they would need an Arctic technique and a knowledge of Arctic conditions'.[21] The policy to fund the scientific exploration of the Arctic, in this instance, had its pay out in the ability to use the knowledge to pursue security policy objectives.

During this period, the US changed their position on the sovereignty of Greenland by surrendering any claims to the island. This became an issue of both security and cooperation as Greenland and Newfoundland could be used as staging points for a Nazi invasion of North America. So, the State Department negotiated the *US-Danish Agreement on Greenland 1941*, where the US confirmed their acknowledgement of the sovereignty of Denmark over Greenland, but assumed 'responsibility of assisting Greenland in the maintenance of its present status', securing 'the right to construct, maintain and operate such landing fields, seaplane facilities and radio and meteorological installations as may be necessary for the accomplishment of the[se] purposes'.[22] As nuclear materials were lost in the ice, the harmful environmental legacy of this policy is still being learned.

Finally, for the indigenous inhabitants of Alaska, US policy continued to maintain the custodial relationship with the tribes. However, given the struggling conditions of Native American reservations in the lower 48 states, the decision was made not to establish reservations in Alaska. Despite the Fourteenth Amendment to the Constitution providing that all persons born in the US would be citizens, given the protectionist relationship of the federal government with indigenous inhabitants, there was uncertainty over whether this applied to the American natives. Thus, Congress passed the *Indian Citizenship Act of 1924*, declaring, 'all noncitizen Indians … are hereby, declared to be citizens of the United States'.[23] While this act indirectly becomes a part of US Arctic policy, it ultimately caused a conflict in relations between the government and the Alaskan indigenous peoples in later years.

There is much about US Arctic policy that is apparently absent in their single image of Alaska issued in 1937. The image promotes the notion of a settled and agriculturally developed region against the backdrop of conserved natures. In this way, the image shows the change in the objectives of exploration from that of discovery and expansion of territory towards the motives of economic incentives. However, these regional development policies were hijacked by the exigencies of global security and government focus changes from settlement to security. In this, policy decisions focused on what objectives should be promoted through government financing. On one hand, this development focused on farming and infrastructure to be able to exploit the wealth of the Arctic and moving economic capital. However, the second focus forced the turmoil within the community of states as they struggled to transition from the expansionist practices to a global system made smaller by technology, the consolidation of power in a few European states and the gradual solidifying of the modern international system. Despite the domestic focus on territorial development, the performances of sovereignty in the next period enhanced the security concerns of the US in the Arctic in the Cold War era.

US Arctic policy in the Cold War

US engagement with the Arctic during the Cold War developed according to the power construct of the international system and the shared understandings of the hostility, suspicion and threats of the bi-polar world. In the Arctic, US policy began to develop a sharper distinction between domestic and foreign policy, although the boundaries between the policies' areas are sometimes as blurred as the performative boundaries between sovereignty over land and sea. As Alaska and the Soviet Union are separated in the Bering Strait by only 53 miles, the reality that 'there is no physical frontier between the United States and the Soviet Union' became significant with the development of aeroplanes and other advanced weapons in the Interwar Years which caused these miles to become increasingly diminutive as a frontier buffer zone.[24] The US tended to rely on bilateral forms of cooperation and security interests dominated the visible engagement with the resources and development of the region. To this end, territorial borders in the Arctic were secured by surveillance technology, and infrastructure was developed to provide energy security while the environment issues were used as neutral, non-security related points of cooperation.

The images and messages in US Arctic postage stamps for this period correspond with the overarching policy approach of the government. Early in the period, two images are produced: one issued on the day of Alaskan statehood in 1959, celebrating the complete incorporation of the territory and its citizens under the sovereign authority of the US, with statehood as a manifestation of the social contract and performance of perfect sovereignty. The second image celebrates the role of the US in exploration of the Arctic and their position with 'first discovery' over the North Pole. This image shows two historical moments: Peary's acquisition of the North Pole and the then recent successful navigation of the *USS Nautilus*, a naval submarine which surfaced through the sea ice at the North Pole in late 1958. It celebrates 'the conquest of the Arctic area by land and by sea'.[25] As the military insecurities of the Cold War and the corresponding arms race between the West and the Soviet Union escalated, this image was a powerful reminder of the superior capabilities of the US to defend their territory and interests – even to the far extents of the icy Arctic. Nearer the end of the period, the dominant messages of postage stamp images would shift.

During the Cold War, the Arctic was less of a scene of discovery and more of a site of reconnaissance. US military planners speculated that if the US were ever attacked, the enemy 'will surely come from over the Pole'.[26] Due to the potential of the Arctic being employed as both a theatre of war and as a route of attack, security interests dominated Arctic policy and funding was channelled towards R&D of technology for Arctic capable operations, especially in the form of submarines, but also including icebreakers and radar surveillance systems. The strategic importance off Arctic military readiness is seen in the commendations of President Eisenhower to the *USS Nautilus* crew who broke through the surface of the ice at the North Pole for their 'revolution in naval tactics and construction'.[27]

Though part of general Cold War policy, the National Security Decision Memorandum 1971 states the policy objective of 'improving the US capability to inhabit and operate in the Arctic'.[28] This competency included development of the Wind-class icebreaker and a 1988 agreement between the US and Canada 'to increase their knowledge of the marine environment of the Arctic through research conducted during icebreaker voyages, and their shared interest in safe, effective icebreaker navigation'.[29] Meanwhile, the US maintained their policy towards the freedom of the seas and the status of the Northwest Passage as an international strait, although they made concessions to Canadian authority on the basis of environmental protection.

The US continued promoting their security interests in the Arctic in cooperation with Canada and Denmark, a legacy of the Interwar Years. In 1951, the US renewed their agreement on Greenland defence 'in accordance with the principles of self-help and mutual aid'.[30] This time, it was to defend against incursions by the Soviet Union, a former ally in the Arctic. Although the Soviet Union had more vulnerable targets in their Arctic than either Canada or the US, the Arctic became a buffer zone for preventing attacks on North America. As part of this strategy, advanced technology systems such as the Distant Early Warning line, stretching from Alaska to Greenland, were installed to prevent surprise attacks from coming from across the Arctic. A security secret labelled *Project 572*, the warning line was 'the most important link in the 10,000 miles outer ring of radar fence'.[31] Built through bilateral US-Canadian cooperation for the purposes of air surveillance and defence, the project and 'NORAD', established in the *North American Air Defense Command Agreement 1958*, formed a significant component of US strategic policy for the region. Demonstrations of the defence system's capabilities include the annual tracking of Santa's sleigh journey from the North Pole every Christmas Eve, an interesting manipulation of sovereignty performances!

Although Cold War presidents were also heavily focused on the development of energy resources, this is heavily contrasted against the background of conservation and environmental concerns. In this period, the US had a second strand of Arctic policy focusing on the emphasis of environmental protection in a two-pronged approach. This is also present in visual representations of policy found in postage stamp images that depict both domestic environmental policies and international cooperation on environmental issues. The environmental aspects of policy appear in images of polar bears and in a 1971 image showing Mt McKinley (now Denali) as a national park, as part of US conservation policy. Although there is a specific policy not to repeat subjects on stamps within the space of 50 years, the polar bear appears on stamps in 1971, 1981, 1999 and 2009.

The first approach implemented the Arctic policy directive of 'promoting mutual beneficial international cooperation in the Arctic'[32] and was pursued through multilateral and bilateral environmental agreements such as the 1973 *Agreement on the Conservation of Polar Bears* with the entirety of the circumpolar Arctic, continuing the policy of the protection of endangered species seen earlier with the protection of fur seals. Including a regional focus on the Arctic, the US agreed to the *US*-Soviet *Cooperation in Environmental Protection 1972* to solve 'the most important aspects of the problems of the environment ... to develop the basis for controlling the impact of human activities on nature'.[33] This policy framework paved the way for the later cooperation emerging in the 1991 *Arctic Environmental Protection Strategy*, signed by all eight of the circumpolar Arctic states.

The second approach of US policy on the Arctic environment is guided by the principle of 'supporting sound and rational development in the

Arctic region, while minimizing the effects on environment'.[34] This aim is promoted by the Arctic Research Commission founded in the *Arctic Research and Policy Act of 1984*, establishing a research agenda to 'mitigate the adverse consequences of development to land', to 'enhance the lives of Arctic residents, [and] increase opportunities for international cooperation'.[35] Development in the Alaskan Arctic includes the extraction of fossil fuels, such as those in Prudhoe Bay and the building of the Trans-Alaskan Pipeline to improve national energy security against the backdrop of global oil crises. This oil travels across the American Arctic, transported from the north to the warm water seaports of southern Alaska where it is loaded onto tankers and so mitigation of environmental consequences through policy mechanism is a critical approach. The failures of this strategy are seen in the 1989 *Exxon Valdez* accident, when millions of barrels of oil spilled into Arctic waters, destroying fishing industries and the local ecosystem. Through this, it is seen that US policy towards national security and economic development supersedes its policy merely to 'minimise effects' on the Arctic environment.

However, despite the turn to multilateral cooperation and policy supporting the domestic concerns on environmental issues, it is in this period that the US developed its position on the possibilities for an 'Arctic Treaty'. Under the Nixon administration, this potential Arctic Treaty System was called the 'Northlands Compact' approach. In a national security memorandum, Secretary of State Henry Kissinger informed the US Departments of State and Defense, 'The President does not desire, however, that the United States undertake discussions at this time with … countries with Arctic interests with the aim of promoting the establishment of a multinational Northlands and Arctic Cooperation Compact....'[36] The position of the US regarding an Arctic Treaty remains unchanged, signalling a policy preference for maintaining the Arctic as a national preserve of strategic resources rather than as a global commons.

Although Alaska finally became a US state in 1959, the exigencies of the Interwar Years and Cold War era policies caused the federal government to retain ownership over more than half of all land in Alaska. Thus, US Arctic policy during the Cold War saw the continuation of similar objectives for the US government, especially in areas of national defence, freedom of the seas and mutually beneficial international cooperation. New objectives introduced included the approaches of environmental and development policy seeing a transition from research to promote national defence towards research in the aim of Arctic resource development for economic and energy security. This was supported by the agencies guiding the policy directions and orchestrated Arctic research that furthered grander US policy agendas. Finally, in a move away from the development of international institutions which are characteristic of the international system in this period, the US positioned itself against an Arctic Treaty System, instead preferring a collection of agreements to organise cooperation in the region.

Contemporary US Arctic policy

US engagement with the Arctic in the post-Cold War period responds to a change in the structure of the international system due to two ideological transitions: the failure of the Communist project behind the iron curtain resulting in the end of the bi-polar system and the global response to terrorism. During this period, the US released three Arctic policy documents under the Clinton, George W. Bush, Obama administrations and although a major policy document has not yet been released, the Arctic is a consideration of the Trump administration. The policies focus on the changing nature of the international system, national security and defence needs including energy development, the strengthening of cooperation between the Arctic states, protecting the Arctic environment and increasing scientific monitoring.

With climate change and environmental concerns moving centre stage in international fora, at least at a superficial level, the US is keen to show its interest in representing itself as a good steward of the Arctic environment. This is with great caveat as environmental protection is only a concern to the extent that it does not interfere with security concerns or the interests of private capital. This is reflected in the policy discourse in which there is a significant amount of tension over the development of energy resources in the ANWR and meanwhile, the US participates in Arctic Council discussions on environmental protection for the region. Corresponding with these trends, the majority of US postage stamps with Arctic images commemorate flora, fauna and environmental integrity, with the polar bear as a reoccurring theme. Almost all visual representation in postage stamps focus on the domestic aspects of the Arctic environment.

Given the political utility of cooperation on environmental matters in the Arctic during the post-Cold War period, the promotion of this characteristic is not surprising. Contemporary international collaboration began in the 1991 *Arctic Environmental Protection Strategy*, a non-binding agreement seeking 'cooperation in scientific research to specify sources, pathways, links and effects of pollution ...'.[37] The negotiation background to this agreement led to the US seeing the Arctic Council as the solution to managing the unique issues of the Arctic environment not already adequately addressed within international law. Before the potential of the Arctic Council was fully realised, some international cooperation on the environment developed external to this body. This included the 1997 Arctic Military Environmental Program, developed to address Soviet Cold War nuclear waste deposits in the Arctic maritime,[38] a programme that continues to clean up the Russian Arctic.

Although US sovereignty in the Arctic is rooted in territorial Alaska, in contemporary policy, 'the Arctic is primarily a maritime domain'; this has the effect of transferring the geographic focus of Arctic policy from the Alaskan territory to the extra-territorial spaces in the Arctic Ocean.[39] Despite Congressional tensions over drilling for oil in Alaska, US official discourse promotes 'environmentally sustainable development' while 'conserving the region's rich and unique biological resources'.[40] Thus, there has been open criticism of Japanese and Canadian whaling practices, specifically declaring Canadian practices as 'unacceptable', because 'Canada's conduct jeopardizes the international effort that has allowed whale stocks to begin to recover from the devastating effects of historic whaling'.[41] Both of these issues concern US national interests in the protection of the Arctic ecosystem, including fisheries, which are impacted by environmental change. US policy formally promotes the pursuit of 'responsible Arctic region stewardship' although the US reputation on the environment is plummeting under the Trump administration.[42]

A multitude of US diplomatic statements continuously reiterate the intent to abide by international law, although policy dictates that it 'is prepared to operate either independently or with other states to safeguard [US] interests'.[43] This statement on respecting international law reflects that the US have not yet ratified *UNCLOS 1982*, which provides for extended territorial claims along the continental shelf. However, territorial claims of up to 200 nautical miles entered customary law after the *Truman Proclamation* in 1945, a principle that was codified into international law in the *Convention on the Continental Shelf 1958*. Thus, it is the position of the US to exercise 'authority in accordance with lawful claims of United States sovereignty, sovereign rights and jurisdiction in the Arctic region'.[44]

Although the US have two realms for sovereignty practices, both in Alaska and in the Arctic maritime, the US holds that except for *extended* continental shelf claims, sovereignty over Arctic territory was confirmed and established in previous eras. Reflecting this view of Arctic sovereignty throughout the circumpolar Arctic, the US responded to Canadian Prime Minister Stephen Harper on his statement that Canada is 'fully committed to strengthening its Arctic sovereignty on every level' with the position that 'the United States does not question Canadian sovereignty over its Arctic islands'.[45] From the US position, there are few questions in the Arctic with regards to sovereign territorial claims.

The primary line of effort in US Arctic policy continues to pursue security interests, especially the interests of defence and energy security. The US policy documents of this era have signalled several different defence approaches with the Clinton policy having the objective of 'meeting post-Cold War national security and defense needs' and the US 'must maintain the ability to protect against attack across the Arctic'.[46] The George W. Bush era policy shows a transition in the national security agenda, concerned with 'fundamental homeland security interests in preventing terrorist attacks ... that could increase United States vulnerability to terrorism in the Arctic region'.[47] US Arctic policy becomes more extensive under the Obama administration seeking to 'advance US security interests by evolving infrastructural capabilities, and the enhancement of domain awareness of activities, conditions, and trends in the Arctic region that may affect our safety, security, environmental or commercial interests' as there are 'many nations across the world [who] aspire to expand their role in the Arctic'.[48] Unfortunately, although the Trump administration has made some movements towards improving Coast Guard capacity, it has rolled back Obama era environmental protection in Alaska and is even gleeful about opening up the ANWR for oil development.[49] To improve the ability of the US to gain recognition of its legitimacy in Arctic fora, Congress designated an Arctic Ambassador from the US Department of State in 2014.

Reflecting changes within the international system with the end of the Cold War, US Arctic policy in this latest period saw the development of several objectives including a transition from the realities of bi-polarity to the need for homeland security and concentration on the need for energy security. US policy also demonstrates awareness that competition for resources in the Arctic requires active development of infrastructure, maritime sovereignty and international law to maintain the Arctic as an exclusive preserve for the Arctic states. To this end, international cooperation has developed from the language of 'minimizing adverse effects to the environment'[50] to 'responsible stewardship [which] requires active conservation of resources, balanced management, and the application of scientific *and* traditional knowledge'.[51] This cooperation increasingly involves multilateral agreements through all eight of the Arctic states on environmental issues. While this has been a strong strategy historically, recent Executive and legislative persuasions make this a difficult position to maintain.

Conclusions

The visual representation of US policy in its postage stamps gives an interesting overview of the narrative of the US relationship with the region, reflecting the dynamics of the US performances of Arctic sovereignty. The vast and frozen wastelands of Alaska, which Seward purchased in an act of imperial impulse, has been one of the greatest sources of territorial wealth in which the US have invested. Since the time of this purchase, the US have had several main policy objectives which include sovereignty and security, the environment, economic development, advancement of scientific knowledge and international cooperation. The images show that the US identifies itself as a responsible environmental steward, even if this image is, in reality, overwhelmed by economic and security issues.

In this long-term overview, US Arctic policy reveals itself to be a complex and multi-layered concern. In many ways, the pursuit of policy objectives and national interests reflects the structure of the international system, as the US have adapted their behaviour to the rules of the expanding community of states over the course of the decades. This is seen especially in the images showing both themes of polar exploration and military defence. It seems that policy approaches are also heavily contingent on the zeitgeist of the age – from imperial expansionism to lack of environmental concern. It is unfortunate that the government has never stopped seeing the American Arctic as a storehouse of economic and resource development.

US engagement in the Arctic has continuously pursued national interests and not merely interests isolated to the Arctic region. This has included broad approaches in the areas of defence, economic security and insistence upon the freedom of the seas, when the US could have claimed the North Pole or applied the sector principle as did other Arctic littoral states. Practices such as this avoidance of the sectoral concept, the introduction of the concept of the continental shelf in the *Truman Proclamation* and the US avoidance of an Arctic Treaty System have demonstrated that as a leading power, the US have been instrumental in shaping the structure of the international system in which Arctic geopolitics occur. And finally, through continued and deliberate pursuit of international cooperation, and restraint from continuing the expansionist practices of dominant powers, US practice in the Arctic acts as a stabilising force in Arctic politics as their foreign policy reinforces and guides the expectations of the behaviour of the other Arctic states.

Notes

1 Corine Wood-Donnelly, 'Exploring and Mapping Alaska: The Russian American Era, 1741–1867', *Terrae Incognitae* 49, no. 1 (2017): 99–101, p. 100.
2 John O'Sullivan, 'Annexation', *United States Magazine and Democratic Review*, 1945, p. 6.

3 James Monroe, 'Monroe Doctrine' (The Avalon Project, 2 December 1823), http://avalon.law.yale.edu/19th_century/monroe.asp.
4 William H Seward, *Speech of William H. Seward at Sitka, August 12, 1869* (Washington DC: Philip & Solomons, 1869), p. 24.
5 Charles Hallock, *Our New Alaska or The Seward Purchase Vindicated* (New York: Forest and Stream Publishing, 1886), p. ii.
6 Molly Lee, 'The Alaska Commercial Company: The Formative Years', *The Pacific Northwest Quarterly* 89, no. 2 (1998): 59–64, p. 63.
7 'The Case for Alaskan Statehood', *Fairbanks Daily News-Miner*, 13 May 1958.
8 James C. Zabriskie, *The Public Land Laws of the United States 1796–1869* (San Francisco: H.H. Bancroft and Company, 1870), p. 874.
9 'Reports on the Committee on Coinage', *New York Times*, 22 March 1880.
10 Ralph S. Tarr, 'The Alaskan Problem', *The North American Review* 195, no. 674 (1912): 40–55, p. 43.
11 Tarr, p. 51.
12 J. Stanley Brown, 'Fur Seals and the Bering Sea Arbitration', *Journal of American Geographical Society of New York* 26, no. 1 (1894): 326–72. pp. 364–5.
13 Isaac I. Hayes, *The Progress of Arctic Discovery* (New York: American Geographical and Statistical Society, 1868).
14 Sara Whitney, 'History of Alaska Population Settlement' (Juneau: Alaska Department of Labor and Workforce Development, 2013), p. 9.
15 James R. Shortridge, 'The Collapse of Frontier Farming in Alaska', *Annals of the Association of American Geographers* 66, no. 4 (1976): 583–604, p. 584.
16 Shortridge, p. 583.
17 Clarence C. Hulley, 'A Historical Survey of the Matanuska Valley Settlement in Alaska', *The Pacific Northwest Quarterly* 40, no. 4 (1949): 327–40, p. 327.
18 Woodrow Wilson, 'First Annual Message December 2, 1913' (The American Presidency Project, 1913), www.presidency.ucsb.edu/ws/index.php?pid=29554.
19 Frank Debenham, 'The Aims of Polar Exploration', in *The Polar Book*, ed. Louis C. Bernacchi (London: E. Allom & Co, 1930).
20 Richard G. Bucksar, 'The Alaska Highway: Background to Decision', *Arctic* 21, no. 4 (1968): 215–22.
21 Vilhjalmur Stefansson, *Arctic Manual* (New York: Macmillan Company, 1944), p. 25.
22 Department of State United States Government, 'Denmark–United States: Agreement Relating to the Defense of Greenland', *The American Journal of International Law* 35, no. 3 (1941): 129–34, p. 130.
23 United States Government, 'Indian Affairs: Laws and Treaties' (United States Government Printing Office, 1929), p. 1165.
24 William Henry Chamberlin, 'The Cold War: A Balance Sheet', *Russian Review* 9, no. 2 (1950): 79–86, p. 80.
25 Post Office Division of Stamps United States Government, *Postage Stamps of the United States: An Illustrated Description of All United States Postage and Special Service Stamps Issued by the Post Office Department from July 1, 1847, to December 31, 1959* (Washington DC: United States Government Printing Office, 1960), p. 170.
26 William M. Leary and Leonard A. LeShack, *Project Coldfeet: A Secret Mission to a Soviet Ice Station* (Anapolis, MD: Naval Institute Press, 1996).

27 Dwight D. Eisenhower, 'Telegram of Commendation to Commander and Crew of the USS Nautilus' (The American Presidency Project, 17 January 1958), www.presidency.ucsb.edu/ws/index.php?pid=11107.

28 National Security Council United States Government, 'National Security Decision Memorandum 144' (United States Government Printing Office, 1971).

29 United Nations, 'Agreement Between the Government of Canada and the Government of the United States of America on Arctic Cooperation' (United Nations, 11 January 1988), https://treaties.un.org/doc/publication/unts/volume%201852/volume-1852-i-31529-english.pdf.

30 Department of State United States Government, 'Defense of Greenland: Agreement Between the United States and the Kingdom of Denmark', in *American Foreign Policy 1950–1955*, vol. I and II, Basic Documents (Washington DC: United States Government Printing Office, 1957).

31 Richard Morenus, *DEW Line: Distant Early Warning* (New York: Rand McNally & Company, 1957), p. 49.

32 United States Government, 'National Security Decision Memorandum 144'.

33 Department of State United States Government, 'Agreement on Cooperation in Environmental Protection' (United States Government Printing Office, 23 May 1972), www.epa.gov/sites/production/files/2014-04/documents/russia-envagreement-1972.pdf.

34 United States Government, 'United States Arctic Policy' (United States Government Printing Office, 1983).

35 United States Government, 'Arctic Research and Policy Act of 1984 (Amended 1990)' (United States Government Printing Office, 1990).

36 National Security Council United States Government, 'National Security Decision Memorandum 202' (United States Government Printing Office, 1972).

37 Arctic Council, 'Arctic Environmental Protection Strategy' (Arctic Council, 1991), http://library.arcticportal.org/1542/1/artic_environment.pdf, p. 2.

38 William J. Clinton, 'Joint Statement on Norway–United States Cooperation' (The American Presidency Project, 15 October 1999), www.presidency.ucsb.edu/ws/?pid=56726.

39 United States Government, 'NSPD-66: Arctic Region Policy' (United States Government Printing Office, 2009).

40 United States Government, 'Presidential Decision Directive/NSC-26: The United States Policy on the Arctic and Antarctic Regions' (United States Government Printing Office, 1994).

41 William J. Clinton, 'Message to the Congress on Canadian Whaling Activities' (The American Presidency Project, 10 February 1997), www.presidency.ucsb.edu/ws/?pid=53558.

42 United States Government, 'National Strategy for the Arctic Region' (United States Government Printing Office, 2013).

43 United States Government, 'NSPD-66: Arctic Region Policy'.

44 United States Government, 2009.

45 George W. Bush, 'The President's News Conference with Prime Minister Stephen Harper of Canada and President Felipe de Jesus Calderon Hinojosa of Mexico in Montebello, Canada' (The American Presidency Project, 21 August 2007), www.presidency.ucsb.edu/ws/index.php?pid=75725&st=arctic&st1=.

46 United States Government, 'Presidential Decision Directive/NSC-26: The United States Policy on the Arctic and Antarctic Regions'.

47 United States Government, 'NSPD-66: Arctic Region Policy'.

48 United States Government, 'National Strategy for the Arctic Region'.
49 Donald J. Trump, '919-Remarks During a Cabinet Meeting' (The American Presidency Project, 20 December 2017), www.presidency.ucsb.edu/ws/?pid=129017.
50 United States Government, 'National Security Decision Memorandum 144'.
51 United States Government, 'National Strategy for the Arctic Region'.

3 The Canadian Arctic

Introduction

In its history as a political geographic space, the Canadian Arctic has a curious evolution. The roots of modern Canada, tenuously claiming sovereignty over the soil (and subsoil) of the Canadian Arctic sector, germinated from the seed of an imperialist project; however, Canada is now seeking to achieve perfect sovereignty over the Arctic by remedying the legacies of colonial praxis. First the Canadian Arctic was explored, claimed and mapped by representatives for the British Empire with these territorial possessions sometimes challenged by other nations, especially France. The empirical evidence of this exploration history is vast and easily traceable in the cartographies of the Canadian Arctic, with place names such as Baffin Bay, Fort Ross and Victoria Island dotting the map. This history is interesting because, while it was the efforts of Britain to establish the Northwest Passage that led to a complete east to west transit of the North American Arctic, British explorers had not made all the new territorial 'first' discoveries. So, when Great Britain transferred sovereignty over all the Arctic lands islands to Canada in 1880, it was not certain that this sovereignty was theirs to give. It is this transfer of imperfect sovereignty that is likely the source of the Canadian crisis of sovereignty over the Arctic, though there are few external challenges to their authority over this territory.

In their quest for satisfying their need of perfect title over the Canadian Arctic, Canada has performed this sovereignty in observable physical acts and in textual and spoken speech acts. Many of these performances of sovereignty are visible in their postage stamps featuring the region, and from these images emerges a visual narrative of Canadian sovereignty over the Arctic. These visual representations of the semiotics of sovereignty seek to communicate the evolution of the Canadian Arctic from discovery through to effective occupation and on to indigenous enfranchisement into the body politic of the Canadian state, as the international system developed in tangent to the political evolution of Canada. They show the different political technologies used in policy to secure this sovereignty, including territorial administration, exploration, development, protection of environmental resources and indigenous inhabitants.

Canada's Arctic engagement in the twenty-first century reveals a state that doubts its own authority over the region; yet is an active participant on the Arctic stage, both in its domestic and foreign policy practices. This relationship is a part of the internal crisis of Canadian Arctic sovereignty where for decades politicians have emphasised the need for Canada to establish or to secure their sovereignty over the Arctic. As a way of framing their sphere of interest, Canada makes claim to sovereignty over the entirety of the lands in the Arctic Archipelago, and in the maritime their claims extend to the geographic North Pole. Canada is one of the states who attempted to establish an Arctic sector, and the failure of this method to take root in international law has left it seeking other methods to establish its sovereignty up to the North Pole. While it seeks to achieve this through other practices, and even though it has no basis in international law, Canada continues to produce official cartographies of the boundaries of the nation using the sector principle to show an international boundary extending all the way to the pole.

Now that new categories of maritime territory have been created in the United Nations Convention on the Law of the Sea (UNCLOS), Canada's claim of sovereignty to the North Pole is challenged by overlapping claims from the other side of the circumpolar Arctic: Russia. As a result, there are sometimes overt performances and brash diplomatic statements made by Canada to their sovereignty over the pole, a claim which confronts Russia's similar assertions to ownership of this imaginary point. Canada's performances of sovereignty over the Arctic show the evolution of the international system, from imperial practices through to reasonable effective occupation of the region and now on to policies including indigenous participation, which could be considered as progressive in the asymmetric international system. In branding Canada as an Arctic nation, their policy methods show they have the approach of layering new practices for performing sovereignty over old practices of earlier times as the international system continues to evolve.

Canadian Arctic policy in the Age of Discovery

Canada's engagement with the Arctic began in 1870 when Great Britain transferred sovereignty over the Northwest Territories and Rupert's Land to the newly created Dominion of Canada. A mere decade later, in

1880, the British Crown then transferred title of its Arctic islands to Canada. This included the British territories acquired by right of discovery from the ventures of Martin Frobisher, Henry Hudson and Sir John Franklin; however, it also included islands discovered by explorers representing other states. This is the moment that the uncertainty of Canada's sovereignty over the Arctic entered the narrative of government discourse; it 'remained uncertain, ... the extent of territories granted [and] the status of the islands north of the mainland'.[1] Since this time, Canada has been attempting to establish, or even prove, the legitimacy of its Arctic sovereignty.

Given the rules for the acquisition of territory, it is possible that some islands belonged to the US or to Norway due to 'first' discovery. Acknowledging this, the wording of the transfer document was ambiguous in the hope that the sovereignty status of the uncertain possessions would be forgotten and the inchoate titles would fail. To allow time for this, 'Canada took no steps to govern or incorporate the added territory between 1880 and 1895', an act verifying the lack of Canadian title.[2] From around 1895, the Government of Canada began implementing effective occupation of the territory to demonstrate their sovereign authority. As a performance of sovereignty, commercial operations, such as foreign whaling operations, were informed that 'they were subject to Canadian regulation'.[3] Today, Canada claims 'the legal basis for Canadian sovereignty over these islands rests predominantly on a mix of cession and occupation, to which considerations of self-determination could be added'.[4] This position is reflected in contemporary Canadian Arctic policy, which frequently makes reference to effective occupation by the virtue of Canada's indigenous inhabitants.

During the Age of Discovery, the Dominion of Canada was in an interesting political position located somewhere in between the conditions of internally and externally sovereign, and even the title 'dominion' indicates their subordinance to another political authority. Created in an imperial project with what in contemporary politics would be called 'home government', Canada is the outcome of the atrophying of the British Empire. The visual representation including the Arctic during this period reflects both Canada's position as part of the British Empire and the uncertainty of sovereignty over the Arctic Archipelago. Issued in 1898, the Canada Imperial Stamp displays a map of the global British Empire, quoting 'we hold a vaster empire than has ever been', with the Arctic islands not quite making the frame. This period forms the foundations of Canadian Arctic policy going forward, as it attempts to negotiate the flaws and methods of earlier imperial projects.

Canadian Arctic policy in the Interwar Years

Canada's sovereignty over the Arctic Archipelago was never challenged by the US or Norway, but the attitude of insecurity was already planted. By 1925, Canada had dabbled in various efforts to demonstrate sufficient sovereignty over a vast area of the Arctic from Ellesmere Island in the east to Wrangel Island, located north of Russia, in the west. Some of this activity included sending explorers to survey the territory, the purpose of Vilhjalmur Stefansson's Canadian Arctic Expedition from 1913–18, and a journey across Arctic Canada by Major Burwash from the Mackenzie Delta to Hudson Bay in 1925–6, collecting scientific information 'under orders from the Canadian Government'.[5] Even today, sending scientists to the Polar Regions is considered an acceptable display of government endorsed sovereign activity in fulfilment of effective occupation in regions not fit for human habitation.

Yet, these performances of sovereignty were not sufficient to dissipate the anxiety of the Canadian government regarding questionable entitlement to Arctic territory, and there was internal pressure to clarify their exact Arctic claims. In light of this, Senator Pascal Poirier launched the sector principle in 1907 in a speech, saying 'the time has come for Canada to make a formal declaration of possession of the lands and islands situated in the north of the Dominion, extending to the North Pole'.[6] Poirier identified that Canada could maintain claims to sovereignty in the Arctic via four legitimate methods: through English discoveries, through transfer of French territories to Britain, through occupation of the Hudson Bay and, finally, through the sector principle. Poirier also claimed ownership of the North Pole by virtue of British exploration and discovery.[7]

Although the government never issued an official declaration defining their sovereign territory, the use of the sector principle by Canada received official endorsement in the 1925 Senate Debates by 'laying claim to everything between the longitudinal lines of 61 degrees west and 141 degrees west, extending "right up to the Pole"'.[8] A parliamentary bill requiring scientists and explorers to acquire a licence to operate in the region soon followed.[9] Finally, the Government of Canada communicated these territorial claims by publishing maps showing the international boundaries of Canada extending up to the North Pole. This includes the Atlas of Canada 1957, which shows the political evolution of Canadian territory when 'by 1925 Canada extended to the North Pole'.[10] Although the sector principle never became a rule of the international system, and though it is in contravention of UNCLOS, official Canadian maps as late as 2017 continue to represent the international border to this point.

When Senator Poirier announced the North Pole as a possession of Canada by transfer from Great Britain, the point itself had still not been reached by any explorer. As a result, he was arguing for the use of a 'spheres of influence' title rather than by 'first' discovery. Within just over a year of the Senator's statements, Peary physically reached the North Pole on sea ice in 1909, planting the US flag. The US then 'refused to claim jurisdiction over the pack ice', as claims to sovereignty over the sea were not permitted in the rules of the international system.[11] However, this fact and precedence has not stopped Canada from continuing to make sovereignty claims over the North Pole.

Although the US have never tried to establish sovereignty by use of the sector principle or the regions of attraction, they never denied Canada's right to do so. In fact, as a result of diplomatic correspondence following the over flight of the pole by a dirigible in the early 1920s, the US acknowledged the sovereign rights of Canada within its sector and relinquished any former potential claims to territory discovered by American citizens.[12] This diplomatic exchange, in reality, solved the first question over the legitimacy of Canada's claim to the Arctic islands.

The visual representations of the Arctic in this period continue to focus on the projection of territorial possession in mapping. In 1927, the 'Map of Canada' stamp was issued; it shows the inclusion of Arctic territories, but it describes the nation as extending from the Atlantic to the Pacific, demonstrating that Canada did not yet identify as an Arctic state. A second image depicts Canada's territory and continued relationship with the British Crown, although during this period domestic sovereignty within Canada was fully realised in 1931. However, Britain continued to be responsible for much of Canada's foreign policy until the security fears of the Interwar Years, such as occupation by Germany, forced these responsibilities to be assumed by Canada. However, the immaturity of the state of Canada forced them to consider a new partner to assist in the security of their newly acquired external sovereignty: their neighbours to the south.

Canadian Arctic policy in the Cold War

As Canada faced a new security threat from across the Arctic Ocean, it entered a new phase of insecurity regarding its Arctic sovereignty. At this point, effective occupation through development became important as a part of national policy. During this period, the pressures of national security forced Canada to collaborate with the Americans in the military defence development of their Arctic territories. When it was realised that the Arctic could be used as an attack route and also as a theatre of attack,[13] the network of collaboration between the two nations became even more entangled, but the physical part of the collaboration primarily occurred in Canadian territory and 'all of this activated created powerful incentives for the development of a new federal approach to northern administration'.[14] The strategic turn in Canada's policy towards the north is strongly reflected in the production of visual representations in the postage stamps of the period, especially as the Diefenbaker government moved to improve economic development of the north and incorporate indigenous inhabitants into the Canadian national project.

As the Cold War increased in severity, technology laid bare the vulnerabilities created by the proximity of the US and the USSR across the minuteness of the Bering Strait, 'transforming the region into a focal point of contemporary strategic thinking'.[15] Early in this period, the Americans invested in the development of a winter road to Alaska across Canada. With security infrastructure installation underway, the US military in the Arctic out populated Canadian citizens three to one.[16] In effect, the US was better at effectively occupying the Canadian Arctic, and Canada was increasingly reliant upon the US for its Arctic security and defence. With Canada unable to secure its Arctic territories without the assistance of another government, this situation continued to provide a source of worry for demonstrating effective Canadian sovereignty in the Arctic. However, it was also this activity that gave Canada the 'infrastructure that promised to render northern resource development practical, while the global markets for these resources were forming'.[17]

As part of the Northern Strategy, scientific and technological advances also began again to contribute to the effective occupation of the Arctic. This is evident in the Polar Continental Shelf Project when the Canadian government began to map the entire maritime continental shelf in the Arctic. This policy began in the same year in which *UNCLOS 1958* codified the extension of the continental shelf into international law. During this time, the Diefenbaker government found it 'expedient to initiate a programme of hydrographic, oceanographic, geophysical, and biological studies of the continental shelf as a direct means of asserting Canadian sovereignty in the High Arctic'.[18] It can be argued that the Canadian government introduced these programmes as a means to alleviate the insecurities in their sovereignty brought about through the continued occupation of foreign military personnel and scientists on their soil during this period.

The development of the Arctic is celebrated in two postage stamps of the period. The first, from 1955, is the image of an Inuit in the Arctic maritime, hunting beneath the shadow of an aeroplane flying overhead. The second, perhaps more significant, image is a 1961 release captioned 'Northern Development', which promoted the idea of the Canadian Arctic as being settled with agricultural development by the pioneers who opened the north for settlement and economic progress. However, this image also perpetuated the myth that the Arctic had already become the new west, a frontier for resource development. Although economic development became possible due to the new infrastructure, such as roads and airstrips that had increased access to the region, the opening of the north for resource extraction caused other problems amongst the existing populations.

The second strand of Canadian Arctic policy during this period is a decidedly unfortunate chapter of Canadian history, representing the continued asymmetric relationship between the government and indigenous inhabitants. In the transfer of the rights of territory from Britain, the Inuit again became a casualty of colonialism, assimilated into the sovereignty of the Canadian state. In this transfer, the indigenous populations were again prey to imperial practices as they were incorporated into the project of nation building. Such is the case with the indigenous populations and the state of Canada, a reality that continues to cause consternation for the Canadian government in their Arctic policy.

As Canada modernised and developed its own national identity, it ostracised the indigenous populations, extracting them from their homelands and decimating their traditional ways of life. When indigenous communities adapted to new conditions brought by the development of the north, many Inuit became 'semi-nomadic', centring around these small economic centres.[19] This caused a problem for the Canadian government, which was worried that the decline in Inuit mobility caused 'a potential weakening of its claim to Arctic sovereignty'.[20] To mitigate this problem, the Canadian government forced many Inuit to move further north, reorganising them into permanent communities. Yet as Arctic indigenous populations began to experience social hardship due to exposure to the south and poverty due to the decline of the fur trade, it became impossible to ignore the conditions of northern peoples given the federal presence in the north as well as the expanding social welfare system of the Canadian state.

As a remedy, the state responded to this issue with a 'full-scale colonial administration to the territorial north'.[21] In this policy, the government attempted a complete reorganisation of indigenous lifestyles. Under this project of assimilation, the policy objective was to 'bring the original inhabitants of Canada to citizenship as quickly as that can reasonably be accomplished'.[22] Children were forced into residential schools, where they lost the influence of the community elders who once would have passed on Inuit traditions and skills through a natural education, a practice which has caused a crisis in indigenous communities.[23] It was not sufficient that indigenous people lost their traditional territory to the forces of imperialism, they were also required to become assimilated citizens of the state of

Canada by default of their birthplace given government aboriginal policy viewing 'enfranchisement as the final confirmation of citizenship'.[24]

The focus on indigenous enfranchisement into the project of identity building to make Canada an Arctic state features heavily in the visual representations of the period. The government issued several series of postage stamps emphasising Inuit culture and heritage, a practice not repeated with other indigenous peoples of Canada. Part of this is the inclusion of sculptures of an Inuit family on a Christmas postage stamp in 1968 labelling them as 'a Christian people'. In 1977, a series celebrated Inuit hunting practices; in 1978, a series on Inuit travel; a 1979 series on Inuit shelter and finally in 1980, a series on Inuit spirits. The stamps of this period display the sense of otherness embedded into the relationship between the Canadian government and the indigenous populations of the Arctic, and belie the harsh realities of the situation of northern peoples.

Finally, during this period Canada found another method through which to enforce sovereignty over their sovereignty claims to territory in the Arctic in the *Arctic Waters Pollution Protection Act 1970*. This environmental legislation defined Arctic waters as 'the internal waters of Canada and the waters of the territorial sea of Canada and the exclusive economic zone of Canada',[25] and reinforced the sector line boundaries drawn in Canadian government cartography. This pollution prevention legislation was a declaration of Canadian responsibility to prevent environmental degradation in 'their' Arctic waters. Through this, Canada practises a performance of sovereignty through environmental stewardship over the Northwest Passage, an area of disputed sovereignty within international relations. These semiotics of sovereignty are also reflected in the visual representations of the period in a 1967 stamp featuring the Queen imposed over the Northern Region, a 1972 image of polar bears on sea ice, a 1977 image celebrating the explorer Joseph Bernier (considered to have officially claimed the Arctic islands for Canada), as well as a 1978 series on ice vessels used for navigation in the Northwest Passage.

Contemporary Canadian Arctic policy

In their statement on Canadian Arctic sovereignty 2006, the Government of Canada states that sovereignty is defined not only as ultimate authority over a given territory, but that it is 'increasingly defined in terms of state responsibility' and is 'linked to the maintenance of international security', including the responsibility of stewardship and occupancy.[26] This position is reflected in their *Northern Strategy 2008*[27] and in their *Statement on Canada's Arctic Foreign Policy 2010*, emphasising the four pillars of Canadian Arctic policy: exercising sovereignty, promoting economic and social development, protecting the Arctic environment, and improving and devolving governance. In addition to using government administration, especially with environmental protection, the methods Canada has used to pursue these interests include the promotion of indigenous inhabitants on the national agenda and a rebranding of Canada as a 'northern nation'. These policy themes are evident in the visual representations of the Arctic, with images featuring the environment, natural resources, exploration and a significant quantity of images with Inuit representation.

When Canada ratified the *UNCLOS 1982* in 2003, rather than increasing Canadian confidence in their Arctic sovereignty, it provided another dimension of insecurity. The treaty provides a ten-year period for a state to make any claims on extensions to its continental shelf in the Arctic. The problem for Canada is that the extended territorial allowances under *UNCLOS* are the same territories over which they have claimed to hold sovereignty since 1925 through the sector principle, if not since 1880 by imperial transfer. It appears the Canadian government sees the treaty as a means of gaining international recognition and legitimacy for their sovereignty over the Arctic maritime where other methods have failed. If Canadian sovereignty over the Arctic up to the North Pole was certain, according to the lines drawn on Canadian maps, there would be little need for Canada to again demonstrate legitimate sovereignty over their 'region of attraction'.

Yet within their international engagement in Arctic policy, Canada exhibits an almost hysterical position on their defence of their Arctic sovereignty up to the North Pole. Since the signing of *UNCLOS 1982*, Canada has been in the process of establishing their claims to sovereignty over the continental shelf. Competing with Canada for this geographical marker is the Russian Federation, both states having earlier claimed up to this point through the sectoral proclamations. Although it remains to be determined through the Commission on the Continental Shelf who, if anyone, can claim the North Pole on the seabed, Canada has embarked on a volley of activity to signify that the North Pole is Canadian. This includes an overtly symbolic performance of sovereignty and state authority in the issuing of Canadian passports to Santa and Mrs Claus in 2012. This act seeks to indicate that as they are Canadian citizens and residents, just as the Inuit are Canadian citizens, the North Pole is within Canadian territory.

However, Canada faces legitimacy challenges to its claims to Arctic sovereignty from another angle: the advancement of the rights of indigenous persons, including the right to self-determination. The entitlement to these inherent rights combined with the negative historical relationship between the Canadian government and the Inuit generates the threat that the Inuit will claim their historical territorial rights to the Arctic and declare political independence from Canada. Thus, the project of enfranchisement and the transforming of indigenous people into Canadian citizens is even more crucial in contemporary policy given the economic value of the region. For the Canadian government which claims sovereignty and territorial rights over the Arctic, the people who have called the Arctic home for a millennium – the Inuit – are the weak point in Canadian Arctic sovereignty as in reality it faces no real external threats from other states within the international system.

The awakening of the importance of the people of the north has been an acutely slow process for the Canadian government. To demonstrate the sovereignty of Canada over its sector of the circumpolar north, the government realises it needs consensus from the peoples of the Arctic. This is demonstrated in the statement from the Inuit Senator Hon. Willie Adams in the Canadian Parliament, saying, 'If we want to assert Arctic sovereignty, we need to ensure that the people from the community are involved. We have been living up there for thousands of years.'[28] This representative of the Inuit people of Canada was communicating to the Canadian government the need for engagement with the Inuit in Canada's demonstration of Arctic sovereignty, perhaps beyond Canadian use of Inuit Rangers to patrol the high north.

The Inuit leaders of today are eager to communicate that they are empowering themselves through education and involvement in the political agendas of the Canadian government to improve the lifestyles of the Inuit, regenerating them as capable leaders of their own affairs. This empowerment is acceptable to the Canadian government for two reasons: the first is that if the Inuit can effectively lead themselves, it results in less cost for the Canadian government, which is heavily subsidising Inuit communities as recompense for past sins. The second reason is that an effective Inuit government working in tangent with the Canadian government provides reinforcement for Canadian claims to the Inuit Arctic for purposes of Arctic sovereignty. This message of the Inuit role has been emphasised from the Inuit Circumpolar Council in a statement that 'an agenda of Arctic sovereignty must involve coordinated strategies to ensure the Arctic has viable and healthy communities, sound civil administration, and responsible environmental management, not just ports, training facilities, and military exercises'.[29]

The Inuit Nation of Canada finds itself in a distinctive position. It is the acceptance of the Inuit to the administrative efforts of the Government of Canada that places Canada in a position to claim Arctic sovereignty on a new basis, that of historic title from the Inuit who have made the Arctic their homeland since time immemorial. This has resulted in many progressive policies for the Inuit, including the formal establishment of their own territory under a consensus form of government and the *2007 Land Claims Agreement*, establishing the Inuit legal right to the resources of their historic land. But there was serious strategy at work behind the development of the devolving governance of the Inuit People when it was claimed in a Senate reading of the Nunivak Inuit Land Claims Agreement Bill, that this land is 'a vital part of our national heritage that our government is determined to protect as we continue to assert Canada's sovereignty in the Arctic'.[30] The territory framed in land claims agreements is even now visible on Google maps, while the international boundary following the sector line is absent.

There are other examples that illustrate the political shift in the gradual awakening to the existence of the north as an integral part of Canada. One such visible example is a reflection on the change of the representation of Canada in official emblems, such as Canada's coat of arms. Canada's arms was initially designed as *A Mari Usque ad Mare*. This phrase has evolved informally by government officials' colloquial use to describe Canada's realm as 'from sea to sea to sea' to include the Arctic Ocean with as much significance as the frames provided by the Atlantic and Pacific Oceans.[31] This audible performance of sovereignty describes the formal incorporation of northern territories into the government administration. This proved that final clause necessary under the customary law of territorial acquisition: effective administrative of government and application of ministerial functions over a people group in the acquired territory. Before this, the Inuit people living within the area of Nunavut had remained largely outside the control or jurisdiction of the Canadian government, which could be perceived as a lack of Canadian effective occupation over the Arctic.

Another important policy element is encouragement for Canadian citizens to embrace the history of the Inuit as their own, and in turn to promote the role of the Inuit as Canadians, with the result that Canada claims historic sovereignty of the Arctic – not through British exploration, but through Inuit heritage. One example of this is found in the Canadian Museum of Civilization in Ottawa. As the Canadian government endeavours to form an attachment of the Canadian people to their Arctic identity, they have gone to impressive lengths to build an iconic temple of Arctic history. The museum offers Canadians an initiation into their national identity when it is at once obvious that the most important aspect of Canadian identity on display is its indigenous people.[32] Very little is said of the imperial and European history of Canada; instead the museum tells a narrative of the Canadian people's history beginning in the Paleo-Eskimo period, an era branded as 'time immemorial'. This is a phrase frequently repeated in government discourse regarding the Arctic, and this narrative reveals Canadian government policy to use the Inuit to maintain their Arctic sovereignty when earlier approaches have failed to diminish their insecurity.

The other facet of this intensified stream is the government continuously emphasising the 'symbolic significance of the north in Canadian national identity'.[33] Some examples of this can be seen in speeches made by former Prime Minister Stephen Harper, where he states, 'Canada's Arctic is central to our identity as a northern nation'[34] and, 'We are a northern country. Canadians are deeply influenced by the vast expanse of our Arctic and its history and legends.'[35] The Arctic has reached a high position on the agenda of the Canadian government due to its importance as a frontier of economic and resource development, and they desire to articulate and even convince Canadian citizens of the value of the Arctic – certainly in economic terms, but most definitely in cultural terms. This was also evidenced in the use of the Inuit Inuksuk as the official emblem of the Vancouver 2010 Winter Olympics, a location thousands of miles away from the place where this figure has any real symbolic or practical use.

The formation of a notion of identity and especially that of a national identity requires the meshing of cultural and social notions of community with political symbols in order to attach the emotional sentiments and psychological processes associated with belonging. Commonly understood national symbols include state flags and national anthems, which are 'carefully constructed, carefully represented projected image[s] of identity that result from a conscious decision-making process'.[36] Indeed, even the Canadian national anthem radiates the realm of the Arctic in the identity of the Canadians. Canada is the 'True North, strong and free'. Not merely part of the north, but a strong, free and true-blue Arctic state.

The importance of the Inuit role in Canadian Arctic policy is emphasised in their postage stamps of the period. The majority of the objects produced in this period contain some representation of the significance of the Inuit to the Canadian Arctic. Many of the stamps combine either territorial administration mechanisms in tangent with Inuit symbols such as the 1999 image of the Northwest Territory with an indigenous musician and the 1995 series of Arctic animals pictured with scenes from northern communities. Other stamps illustrate territorial administration and the environment, such as the 2002 image of the northern lights on a stamp of the Northwest Territory or evidence of government administration in the 2006 image of the Canadian flag symbolically waving over ice fields. Given the continued significance of the Canadian strategy to secure sovereignty over the Northwest Passage, the postage stamps celebrating the work of explorers or that of James White, the cartographer who first mapped the

Arctic into the Canadian atlas, are largely unsurprising. However, the most telling postage stamps of all are the visual representation connecting the Inuit to the broader Canadian nation in the two Olympic Games series in 2009 and 2010, both series including the Inuksuk, the symbolic Inuit landmark.

Conclusions

The long-term incorporation of Arctic territories into the Canadian realm, from the foundations of the Dominion of Canada to the publication of their 2010 Statement on Canada's Arctic Foreign Policy, shows an ongoing crisis of legitimacy within Canadian policy over the status of their Arctic sovereignty. Emerging from this ongoing crisis, the Canadian government has attempted several methods for establishing and exercising their sovereignty including claiming the Arctic through imperial transfer, effective occupation, the sectoral principle and, most importantly, through their relationship with the Inuit. In addition to these formal methods, it has used informal methods to promote their state as a sovereign Arctic nation through the creation of an Arctic identity. This includes the rigorous promotion of Canada as an Arctic nation, or a northern state, with gratuitous use of Inuit symbols and other performances of sovereignty through the use of state administrative technologies.

The subject of Arctic sovereignty has received a significant amount of attention from Canadian policy makers in recent years, and many government officials frequently express the need to establish and/or reinforce sovereignty in the Arctic. Yet, aside from the symbolic claims to the North Pole, it does not appear that any real external challenges to Canadian Arctic sovereignty exist. The challenges are primarily born within, created through a sense of political insecurity due to the dubious imperial transfer, use of the moot sector principle and the subsequent relationship with the indigenous inhabitants. Because of this crisis of legitimacy, Canada has promoted the role of the Inuit as the key to their security of Arctic sovereignty, placing significant effort into incorporating the Inuit into the body politic of Canada. This identity is needed to give the government legitimacy and popular support for the pursuit of its Arctic policies on both the national and international stage, and is being pursued through devolution of government administration to the Inuit people in an attempt to prove Canadian Arctic sovereignty via Inuit rights.

Although Canadian Arctic policy continuously makes reference to establishing Arctic sovereignty, it appears that from a legal position there is no uncertainty of Canadian Arctic sovereignty. They have a historic claim reaching back more than a century, have continuously represented the Arctic on official documents and have demonstrated effective occupation as much as possible in a hostile climate. Thus, this crisis of legitimacy regarding Arctic sovereignty over *terra firma* is a somewhat irrational situation, possibly reflecting that Canada is not willing to relinquish any claims to the North Pole nor any maritime territory they deem to hold, despite the fact that claims to this territory have not been made according to rules accepted within the international system.

Notes

1 Gordon W. Smith, 'The Transfer of Arctic Territories from Great Britain to Canada in 1880', *Arctic* 14, no. 1 (1961): 53–73, p. 53.

2 W.F. King, 'Report upon the Title of Canada to the Islands North of the Mainland of Canada' (Ottawa: Government Printing Bureau, 1905).

3 William J. Mills, *Exploring Polar Frontiers: A Historical Encyclopedia* (Santa Barbara: ABC-CLIO Inc, 2003), p. 129.

4 Government of Canada, 'The Arctic: Canada's Legal Claims' (Ottawa: Library of Parliament, 2008), www.res.parl.gc.ca/Content/LOP/ResearchPublications/prb0805-e.pdf.

5 L.T. Burwash, 'Across Arctic Canada 1925–1926', *The Geographical Journal* 74, no. 6 (1929): 553–68, p. 553.

6 Canada Senate Debates 1907 in Donat Pharand, *Canada's Arctic Waters in International Law* (Cambridge: Cambridge University Press, 1988), p. 8.

7 Canada Senate Debates 1906 in Ivan L. Head, 'Canadian Claims to Territorial Sovereignty in the Arctic Region', *McGill Law Journal* 9, no. 3 (1963): 200–226, p. 204.

8 John Honderich, *Arctic Imperative: Is Canada Losing the North?* (Toronto: University of Toronto Press, 1998), p. 30.

9 Gustav Smedel, *Acquisition of Sovereignty Over Polar Areas* (Oslo: Kommision Hos Jacob Dybwad, 1931), p. 66.

10 Government of Canada, 'Atlas of Canada: Political Evolution' (Queen's Printer, 1957).

11 Donat Pharand, *The Law of the Sea of the Arctic* (Ottawa: University of Ottawa Press, 1973), p. 169.

12 Paul Fauchille, *Traité de Droit International Public*, vol. 2 (Paris: Rousseau, 1925).

13 V.H. Patriarche, 'The Strategy of the Arctic', *International Affairs* 25, no. 4 (1949): 366–474.

14 Frances Abele, 'Canadian Contradictions: Forty Years of Northern Political Development', *Arctic* 40, no. 4 (1987): 310–20, p. 312.

15 Oran R. Young and Gail Osherenko, *The Age of the Arctic: Hot Conflicts and Cold Realities* (Cambridge: Cambridge University Press, 1989), p. 18.

16 David Judd, 'Seventy-Five Years of Resource Administration in Northern Canada', *Polar Record* 14, no. 92 (1969): 791–806.

17 Abele, 'Canadian Contradictions: Forty Years of Northern Political Development', p. 312.

18 Richard C. Powell, ' "The Rigours of An Arctic Experiment": The Precarious Authority of Field Practices in the Canadian High Arctic, 1958–1970',

Environment and Planning A: Economy and Space 39, no. 8 (2007): 1794–1811, p. 1801.

19 James S. Frideres, Madeline A. Kalbach, and Warren E. Kalbach, 'Government Policy and the Spatial Redistribution of Canadas Original Peoples', in *Population Mobility and Indigenous Peoples in Australasia and North America*, eds. Martin Bell and John Taylor (London: Routledge, 2004).

20 Frideres, Kalbach, and Kalbach, p. 100.

21 Abele, 'Canadian Contradictions: Forty Years of Northern Political Development', p. 312.

22 Government of Canada, 'Official Report of Debates' (Ottawa: House of Commons, 1949), p. 2285.

23 Hugh Brody, 'Aboriginal History' (Canada and the Arctic, Canada House, London, 2010).

24 John S. Milloy, *A National Crime: The Canadian Government and the Residential School System* (Manitoba: University of Manitoba Press, 1999), p. 359.

25 Government of Canada, 'Arctic Waters Pollution Prevention Act' (1970). (Government of Canada, 1985).

26 Government of Canada, 'Canadian Arctic Sovereignty' (Library of Parliament, 26 January 2006), www.res.parl.gc.ca/Content/LOP/ResearchPublications/prb0561-e.htm.

27 Although the official languages of Canada are English and French, and thus all government documents must be produced in both languages, the Northern Strategy 2008 is also published in Inuktitut.

28 Government of Canada, 'Canada Senate Debates' (1st Session, 39th Parliament, Ottawa, 7 June 2006). §1510.

29 Inuit Circumpolar Council, 'Arctic Sovereignty Begins with Inuit: Circumpolar Inuit Commit to Development of 'Inuit Declaration on Sovereignty in the Arctic'" (Inuit Circumpolar Council, 10 November 2008), www.inuitcircumpolar.com/uploads/3/0/5/4/30542564/pr-2008-11-10-kuujjuaqsovereigntyworkshop.pdf.

30 Government of Canada, 'Canada Senate Debates' (39th Parliament, 2nd Session, Ottawa, 14 November 2007). §1335.

31 Senior Canadian Diplomatic Personnel, Interview at Canadian Embassy London, 10 December 2010.

32 Canadian Museum of Civilization, *Written in Stone: An Architectural Tour of the Canadian Museum of Civilisation*, Ottawa, Canadian Museum of History, 2004.

33 Klaus Dodds, 'We Are A Northern Country: Stephen Harper and the Canadian Arctic', *Polar Record* 47, no. 2 (2010): 371–82, p. 372.

34 Stephen Harper, 'Prime Minister Stephen Harper Announces New Arctic Offshore Patrol Ships' (Government of Canada, 9 July 2007), www.canada.ca/en/news/archive/2007/07/prime-minister-stephen-harper-announces-new-arctic-offshore-patrol-ships.html.

35 Michaëlle Jean, 'Speech from the Throne to Open the 40th Parliament, 3rd Session, Parliament of Canada' (Opening of Parliament, Ottawa, 3 March 2010), http://publications.gc.ca/collections/collection_2010/bcp-pco/SO1-1-2010-eng.pdf.

36 Karen A. Cerulo, *Identity Designs: The Sights and Sounds of a Nation* (New Brunswick, NJ: Rutgers University Press, 1995), p. 5.

4 The Russian Arctic

Introduction

The Russian Arctic has a three-part history as a political geographic space under the administration of three different political regimes. The first part of this history begins in Imperial Russia, the second part as the Soviet Union and now, the Russian Federation. First, the Russian Arctic was mapped into the empire by explorers and expeditions selected by the Russian Academy of Sciences and the Russian Naval Academy. Second, the Arctic was incorporated into the administrative and economic mechanisms of the state, as the empire sought to fulfil the normative requirements of effective occupation. When the empire was overthrown in the Bolshevik Revolution, the Soviet Union emerged and began a new project of nation-building, and the Arctic became part of this wider mission. With the end of the Cold War and the reshaping of the state into the Russian Federation, the Arctic has again been incorporated into national focus as a strategic resource base. Throughout each of these internal transitions, the relationship of Russia with the Arctic has adapted to the rules of the international system as it pursues its national interest in the region.

As Russia seeks to communicate the legitimacy of its authority over the region, it engages in performances of sovereignty that are visible in its postage stamps. From these images emerge an understanding of how Russia communicates its sovereignty over the Arctic in line with transitions in the international system. These approaches can be seen in the images of the Arctic that narrate the story of Russian exploration, discovery and effective occupation of the region. Part of this account is rooted in the exploration that resulted in the extensive territorial expansion of Imperial Russia across the entire Asian Arctic, the Bering Sea and all the way to Alaska. Another part is anchored in the demonstration of the political technologies that illustrate Russian effective occupation of its Arctic maritime, largely through scientific expeditions and its maritime administrative competencies.

Arctic engagement in the twenty-first century shows the Russian Federation to be engaging with the region by maintaining it as a central part of its broader national strategies, especially where it is seen as a strategic resource base. This relationship with the Arctic is in part due to internal Russian politics where the Arctic continues to be a space where the masculine competencies of the state can be projected, drawing attention away from other domestic issues. Russia is the largest of all Arctic territories in the circumpolar region and, standing to make significant territorial gains from the provisions of UNCLOS 1982, it is keen to use this framework to legitimise its long-standing claims to the region, without losing any territory not covered by these provisions. Russia does this by conducting performances of sovereignty using both old rules of territorial acquisition and newer normative understandings, such as those found in maritime law and in some of the frameworks being developed through the Arctic Council.

While Russia holds a certainty of title over significant terrestrial areas and large swathes of maritime territory in the Arctic, it also has the potential to 'lose' territory at the North Pole. This is an area over which it has made claims and conducted performances of sovereignty towards maintaining over the last century, through military exercises, scientific exploration and administrative infrastructure. Given the national attachment of Russia to the extended Arctic, it continues to make policy and strategic performances that draw international attention to the existence of these long-standing claims. The contemporary policy of Russia towards the Arctic takes in their broader global interests and combines it with their regional attachments. The long-term consideration of Russian engagement with the Arctic demonstrates that their contemporary policy is the accumulation of centuries of sovereignty over the Arctic. In addition, this engagement is influenced by its inherited identity resulting in state practice that sometimes appears to contradict the contemporary rules of the international system as Russia pursues its national interests in the Arctic.

Russian Arctic policy in the Age of Discovery

During the Age of Discovery, Imperial Russia expanded throughout the Arctic by applying the laws of territorial acquisition for the period. By exploring and developing territorial boundaries through unilateral declarations and bilateral treaties, they established sovereignty in their Arctic territories through administration including levying of domestic legislation and incorporating the Arctic into economic development. However, the politics in this period resulted in both the expansion across the Arctic and the eventual contraction of the Russian empire according to the measure of their national interests in the region as they responded to concerns over sovereignty, security and economic issues in the Arctic. The postage stamps of Imperial Russia portrayed only the Tsar or the double headed eagle; as such, there are no visual representations of Russian performances of Arctic sovereignty in this period.

Exploration of the Russian Arctic began in the Middle Ages with the voyages of the Pomors. This marked the start of the significance of the Northern Sea Route (NSR) as part of the 'maritime corridor known as the Great Mangazea Route', beginning at the Kola Peninsula with explorers moving eastward.[1] This route was exceedingly lucrative for the fur trade, the ivory trade and for hunting and fishing resources. Due to the abundant economic resources coming from this area, adventurers came to the region from all over Europe, and Russia could not prevent poaching or the many foreign traders who avoided paying taxes to customs officials. So in 1619, the Russian authorities 'issued a decree forbidding the use of the sea route',[2] stagnating the economic development of the Russian Arctic until the Russian bureaucratic system could administer the territory.

Before exploring the Arctic, Russia initially focused their efforts on imperial expansion southward, and as a result 'Siberia was scarcely known to the Russians before the middle of the sixteenth century'. However, once exploration revealed the wealth of the Sibir region, 'came soldiers, mercenaries and Cossacks, led by Moscow-appointed administrators' who 'found "new lands", built forts and imposed fur tribute'.[3] Russia annexed the Arctic into its empire and established the Siberian Department which was responsible for managing the conquest and administration of the region, especially for collecting tax revenues from the fur trade. As a way of framing its sphere of interest, Russia formally excluded China, its nearest political rival in the east, to any Arctic claims in the *Treaty of Nerchinsk 1689*. It was then organised as an economic region and 'the Muscovite state participated actively in the exploitation of the resources of Siberia'[4] in a linear extraction of resources, an arrangement that would inevitably lead to the unprofitability of the extended Russian Arctic empire.

In the late seventeenth century, Russian military power returned its focus to the sea after the successes on land across the Asian continent, hoping to establish a Russian trading route that would bring additional profitability from the payment of tolls from foreign merchant ships.[5] The Russian Navy commissioned Vitus Bering to lead expeditions to the Arctic. The Tsar instructed Bering to:

> seek the point where it [Kamchatka] connects with America and go to some settlement under European rule, or if any European vessel is seen, learn of it what the coast visited is called, which should be taken

down in writing, an authentic account prepared, placed on the chart and brought back here.[6]

This signalled not only the importance of European discovery for the laws of territorial acquisition, but also of the process of recording any expeditions and discoveries for Russia should they not be otherwise claimed.

Leaving evidence of 'first' discovery, the Russians used a variety of methods for their ceremonies of possession to demonstrate ownership, such as burying plates with seals of the imperial government, erecting totems depicting symbols of the Russian state, and giving sculptures of the imperial crest to native chieftains. These methods 'were designed to be visible warnings to rival nations that the land was under claim. Crosses placed on prominent head lands, highly visible from the water, announced to late-coming seafarers that there was "no vacancy"'.[7] Despite this, the Russian claims to the North American Arctic were challenged by Spanish, British and American overland expansion in the late eighteenth and early nineteenth centuries. To combat this deliberate encroachment, the Russian government proclaimed in 1821 that all commerce and industry in their Arctic spaces was 'exclusively granted to the Russian subjects'; it also prohibited all foreign vessels from approaching or landing on their territories.[8] The intent of this claim was to establish absolute sovereignty over the entire spatial area of the Russian discoveries in the Arctic, both on land and at sea.

Russia implemented effective occupation of Alaska by establishing the Russian-American Company in 1799 and granted the company exclusive economic rights within the territory for a period of 20 years. This resulted in 'an ostensibly commercial company managing territory and populace on behalf of an empire who believe that this arrangement amounted to a cost-effective form of colonialism and provided a convenient scapegoat in case of failure'.[9] In exchange for economic monopoly, the 'government demanded that the company provide social services' to help fulfil the state obligations of effective occupation.[10] However, the combination of corruption, reports of near bankruptcy, and the increasing economic pressures from competitors made it impossible for the colonisation project to retain favour with the government, who was dealing with an array of political pressures in the motherland.

To alleviate some of the pressures of maintaining effective occupation in the Alaskan colonies, the Imperial Russian government negotiated the *Russian-American Treaty 1824* and the *Anglo-Russian Treaty 1825*, establishing the eastern borders of the Russian Arctic against the British North American Arctic. These treaties also had the effect of diminishing Russia's claim of *mare clausum* over the North Pacific, attempted in the ukase of

1821. The treaties made arrangements for navigation, fishing, and trade with natives and explicitly restricted the importation of armaments into the region and the sale of arms to the native populations. The deterioration of this claim to *mare clausum* over the Bering Sea at this critical point in the Age of Discovery prevented sovereignty over this entire sea area from becoming an accepted norm in the international system.

Although the conflicts near to home, notably in the Crimea between Russia and Great Britain, did not result in any change of borders between these parties in North America, there was certainly a realisation for Russia that it was difficult, if not impossible, to defend their territory in Alaska. The colonisation and effective occupation of Alaska by the Russian–American company was subpar at best: and the British had noticed that the Russians had failed to occupy the territories that they had claimed in the 1825 treaty and that 'what Russia has already obtained on this coast [is] by a claim void of any foundation'.[11] When faced with the likelihood of the erosion of their Arctic territories in North America, the Russian government had few options but to sell Alaska to the Americans, simply to avoid losing it to Britain, and so agreed to the *Alaska Purchase Treaty 1867* with the US. This treaty ended Russia's relationship with the Arctic in North America.

Meanwhile, in the Russian Arctic territories of Europe and Asia, exploration and development had stagnated due to a climatic barrier. This was known as the Little Ice Age, a serious cold period of a few centuries that affected the conquest of the Arctic, as the conditions 'prevented foreign ships from sailing along the Northern Sea Route'.[12] When these atmospheric conditions began to decline in the mid to late nineteenth century, foreign interest reignited in the NSR and European states sent expeditions into the Russian Arctic maritime regions. This resulted in questionable sovereignty over places such as Svalbard, Novaya Zemlya and the Wrangel Islands as they were 'discovered' multiple times and utilised at varying points by several states. Russia then realised that a 'way to defend Russian rights of sovereignty in the Arctic was to take part more actively in polar exploration'.[13] To increase their claims by right of the 'most historic' discovery and sufficient effective occupation, interest in the Russian Arctic maritime experienced the 'renewed interest of the Russian government in Arctic Studies',[14] outfitting ships and organising expeditions through the Russian Academy of Sciences.

Security interests were introduced to the Russian Arctic in the 1904 Russo-Japanese War when the fleet based in Russia's eastern warm water port, leased in Port Arthur, was attacked without warning. The war taxed the capabilities of the trans-Siberian railway, which had carried goods from the Pacific to Western Russia. The additional importance of the NSR became obvious, as the war 'showed at a stroke the enormous advantages of the much shorter Arctic Passage' when the rest of the Russian Navy had to sail halfway around the world before it could provide support.[15] Thus 'immediately after the war the government made available funds to explore this passage', including the development of a fleet of icebreakers.[16]

As the Age of Discovery drew to a close, the continued discovery of new islands and the opportunities for scientific discoveries with both economic and security implications caused both insecurities and opportunities for Russia in the Arctic. The increasing geostrategic importance of Russian development along the coast, including ports at both ends of the NSR, elevated the significance of the Arctic. Yet, due to the lack of infrastructural development in Siberia and the lack of political coordination to develop the infrastructure, the NSR would remain relatively quiet for another century.

Russian Arctic policy in the Interwar Years

The Russian relationship with the Arctic faced a dramatic change in the Interwar Years. Not only was there a change in the structure of the international system that resulted from the conflict and cooperation of the Great Wars, but there was also a shift within the domestic political structure of Russia. This occurred during the Russian Revolution, which resulted in the removal of the Tsarist Imperial government that had ruled Russia for centuries and was responsible for past Arctic policy and engagement for Russia. This long-standing political structure was replaced by a new government based to some extent on principles of Marxism, including the removal of the aristocracy and the theoretical equalising of class hierarchies through the redistribution of wealth and property. These international and domestic transformations changed how the Arctic was incorporated into Russian political consciousness, with implications for the development of Arctic policy.

By now, the concept of *mare clausum* had failed to establish control over the vast maritime areas of Arctic Ocean spaces and it was accepted that sovereignty of territory on ice could not be annexed through discovery. Yet, there was widespread belief that undiscovered islands still existed deep within the Arctic maritime, and the states around the Arctic peripheries sent out regular expeditions to discover and map these territories. This included exploration in maritime spaces traditionally considered by Imperial Russia to be located in enclosed Russian seas. These foreign intrusions caused concern for the new Soviet government, who needed to establish their own policy towards Arctic maritime sovereignty, and so 'after a government protest failed to produce the desired results, the

Soviets decided to take action'.[17] The response was to send out their own expeditions, raise the Soviet flag on these territories and arrest foreigners throughout this territory. However, this was not the only threat to Russian sovereignty in the region.

Russian sovereignty in the Arctic was also being challenged by expeditions coming by air, when explorers used zeppelins and aeroplanes to search the Arctic Ocean for unclaimed islands. Having been ravaged by the First World War, a civil war and a famine, Russia had put few resources into developing aviation technology and thus were years behind in developing technology that would enable them to seek new Arctic territories in the same manner. To limit the damage caused by their technological incapacities in their Arctic 'region of interest', the Soviet government issued a declaration in 1926, putting the sector principle into effect. They declared that 'All lands and islands, both discovered and which may be discovered in the future ... located in the Northern Arctic Ocean, ... up to the North Pole ... are proclaimed to be the territory of the USSR.'[18] This declaration established the parameters for the international boundaries printed on Russian maps to this day.

The establishment of the Russian Arctic sector had the immediate result of eliminating territorial insecurities for the Soviet government, but it also produced a new type of responsibility within the realm of effective occupation. This was evident in the 1928 *Airship Italia* crash when the surviving crew needed rescue in the Russian sector. Initially, Stalin decided that Russia would not get involved, but the media had inadvertently committed Russia to the rescue and the government was forced to activate its mothballed icebreakers to prevent loss of face.[19] Although the rescue attempt was the cumulative effort of an internationally coordinated search and rescue effort, it was Russian icebreakers that rescued the survivors. This rescue established Russia's reputation for technological expertise and prowess in Arctic operations and the subsequent media praise 'opened the authorities' eyes to the propaganda possibilities of the Arctic'.[20]

The Soviet project included the notion of 'mastery over nature', and the Arctic constituted 'the ultimate battleground in the Soviets' great "struggle against the elements"' where 'triumphs in the Arctic were translated into victories of almost cosmic significance'.[21] Following the *Italia* incident, icebreakers became a part of the Soviet agenda for northern development as well as part of the myth of the Arctic representing Soviet capabilities to overcome the difficulties of the region. The Stalin-class icebreakers, built between 1932 and 1939, were glorified in both literal and visual arts, with one poet penning the words: 'It is not vernal water overflowing, it is our icebreakers approaching. Not alone is the glorious hero.'[22] In its visual representations, the Soviet government produced many postage stamps with images of icebreakers such as the 1931 *Malygin* issue and a 1940 series celebrating polar drift experiments. These images of icebreakers show their role in supporting scientific research and development in the Arctic.

Even though icebreakers were used to promote the myth of the Arctic, they were truly an important component of Russian Arctic policy during the Interwar Years, due to their supportive role in keeping the ports of the NSR open during the winter. The icebreakers were decisive in the development of the route, as without their support, Arctic shipping would have been impossible.[23] The Soviets invested significant research efforts into developing icebreakers that could keep the ports and the NSR open, bringing coal and other minerals acquired through mining throughout the region. Arctic development also included the establishment of 'refuelling' ports, which increased the range in which icebreakers could operate.[24]

Complementing these icebreakers was the development of Arctic aviation where in addition to supporting scientific exploration and experiments, the aeroplanes acted as aerial scouts for icebreakers. Arctic pilots were given the same honours later given to Soviet cosmonauts, as 'without aviation, the whole scheme of industrial development, settlement, geological and geographical exploration, and of navigation along the NSR would be unthinkable'.[25] Aviation contributed to both the economic policy and security goals of Russia by making it possible for ships to transport minerals and coal mined in the Arctic to the industrial core. They also sighted additional developmental possibilities from air and gave the Soviet military the ability to patrol borders more efficiently. These aviation exploits were commemorated in an extensive 1938 series of postage stamps, celebrating polar overflight by Soviet pilots in stamps that also depicted the Soviet flag flying over the North Pole.

In their land-based Arctic policies, Russia continued to alter the landscape of Arctic peoples, influenced by the overall transformation of Imperial Russia into the Soviet Union. As part of this, the Russian government began to focus more intently on the restructuring of society and the economy in line with the new political ideologies of the nation. These new strategies were implemented in the Five Year Plans, which had severe implications for the Arctic as Soviet style plans were applied to every facet of society, including Arctic indigenous peoples, interfering with traditional lifestyles and economic practices.

Although indigenous lands were already collectively 'owned', the Soviet government began implementing a nationwide policy of collectivisation which intended to eliminate the feudal structure of Russian society and replace it with state ownership of property. When this policy of state collectivisation was applied to indigenous lands throughout the Russian Arctic, indigenous people were assessed as owning too many reindeer per person.

Due to this assessment, thousands of reindeer were confiscated, killed and redistributed amongst collective farms. The result was that many indigenous peoples died from starvation.[26] Although the government made indigenous populations exempt from many 'Soviet duties' such as conscription, the 'state appropriated tribal land and began to cash in on its vast resources – including fish, timber, fur and gold – through industrialization'.[27]

One significant policy managed the economic development of extended Arctic spaces through labour provided by forced migration. Part of this used immigration as a punitive policy, a legacy of the Imperial Russian period. With the Russian Arctic considered as a vast empty space, this migration was necessary to maintain effective occupation and to exploit the economic resources of the region. The Arctic was used to imprison societal undesirables, and although everyone dreaded exile to Siberia, in some places such as the Arctic island of Sakhalin the conditions were so bad that occupants of the penal colony hoped to be moved to Siberia.[28] The NSR was used to transport thousands of 'migrants' to the work gulags throughout Siberia, providing critical labour for the extractive industries that fuelled the Soviet economy.

There is much about Russian Arctic policy in this period that is absent from their visual representations, with forced migrations and indigenous reorganisation clearly omitted from the produced narrative. However, the images present many of the fears of Russia, especially in the challenges to their sovereignty and authority over the extended maritime spaces of the Arctic. All Russian postage stamps show the historic methods for their acquisition of Arctic sovereignty and some of the methods used for maintaining its title. This includes a 1943 image celebrating the exploration and contributions of Vitus Bering to Russian Arctic sovereignty. It also shows the role of scientific exploration in maintaining effective occupation, such as the 1932 image on the International Polar Year. However, the most significant theme is the use of the NSR including the 1935 images of the *Chelyuskin* Expedition and other images of icebreakers throughout the period.

Russian Arctic policy in the Cold War

Russian Arctic policy during the Cold War was partially a continuation of the Arctic policies developed during the Interwar Years, but it also reflects the new structure of the international system, along with its ideological fears and insecurities. In this transition, the continued leadership of Stalin over the Soviet Union provided some continuity in the policies between these periods. Due to the insecurity, suspicion and antagonism of the Cold War, Russian Arctic policy makes a noticeable shift in the Soviet relationship with the Arctic. This relationship reflected the new political realities as they pursued their national interests primarily in security and economic development, only engaging with issues of social development and environmental protection when they factored into the grander Soviet strategy. Near the end of the Cold War period, it was Russian efforts that created the opportunity for future international cooperation in the region.

Dominating Russian Arctic policy during the Cold War period was the combined focus on sovereignty and security in the region. At the beginning of the period, the Soviet Union continued to use the sector principle to demarcate their territorial boundaries. Although the sector principle failed to be accepted as a rule of the international system, its continued use by Russia has several logical bases. First, given the increase in aviation technologies, combined with the increasing attitude of hostility and insecurity, the use of the sector principle gave the Soviets a clearly designated area for understanding sovereign airspace, a practice that continues today. This area is now also used as their formal area of responsibility in maritime search and rescue operations. In this regard, the sector principle is still an integral part of Russian policy.

Second, when sea ice was used in a manner resembling the use of land for scientific research activities, the sector principle framed the areas of territory where its scientists worked. Russia installed dozens of drift stations, many of which were permanently manned by scientists in order to study ice and weather conditions, sea currents and geomagnetic observations to 'ensure the maximum development of the Northern Sea Route', which included making 'sea and air navigation in the Arctic more secure' through 'the systematic collection of all the data which [were] needed to improve ice and weather forecasts'.[29] For many years, Soviet scientists were prohibited from collaborating with other international scientists; the government also restricted foreign research in their territorial waters and refused to let foreign ships use the NSR.[30]

The importance of the NSR within the policy of the period is demonstrated in the efforts to make the passage operable. Not only did the government fund extensive scientific exploration for the entirety of the Cold War, but it also invested in the technology necessary to keep the passage open, and created a bureaucratic mechanism charged with maintaining the administrative functions of the route. In 1956, the NSR Administration managed '15 icebreakers, 100 ocean-going freighters, 150 aircraft, and 35,000 employees … operat[ing] entirely in Arctic waters and handl[ing] up to 2,000,000 tons of freight in a season'.[31] When compared to the capabilities of their Arctic neighbours, it is clear that Russia considered their Arctic developmental operational capabilities of high significance, although this would have been assisted by the Soviet economy, which did not require profitability.

This focus on the operational capabilities of the Arctic maritime has rationale in both military security and economic development. Given the status of the arms race as a predominant feature of the Cold War, and with the unfortunate geography of the Baltic Sea and the Arctic Ocean (Russia's key strategic maritime points, which have ice cover in the winter), this fleet of icebreakers was a significant component of defence policy. Given this strategic importance, 'the Arctic emerged as an arena of naval super-power competition'.[32] Although the Russians did not widely publicise these military capabilities during the height of the Cold War,[33] Arctic operating bases became viable options for Russia and the source of fear for their rivals. This intensive activity would also cause Russia to use the Arctic as a nuclear dumping site, a pollution issue that would provide future opportunities for international cooperation.

Economic development and exploitation of the resources of the Arctic was another significant feature of Russian policy during this period. Supported by NSR infrastructure, this targeted extraction of hydrocarbons, first with coal and then moving onto gas and oil to fuel the expansion of industrial development. Russia also mined other minerals, as this part of the Arctic is 'well-endowed with non-renewable resources, such as nickel, cobalt, platinum, copper, gold, tin, iron and diamonds'.[34] Supporting mining activities, cities were founded in the Russian far north, providing workers for infrastructural growth and transportation networks, such as the Leningrad-Murmansk railroad, used to bring these resources to the populated areas of Russia. This development 'accelerated emplacement of infrastructure that is applicable for both military and economic purposes and the drive to promote colonization, industrialization, transportation and communications links and the overall economic modernization' of the Arctic region.[35]

The totality of the Soviet Communist conception of economic structure included not only the exploitation of natural resources to support the core economy, but also the alteration of the traditional lifestyles and economies of Russian indigenous peoples. This approach resulted in the almost immediate devastation of populations due to famine and starvation, but the social effects of this restructuring compounded when substance abuse and suicide rates increased. The effects of the development policies on indigenous lands resulted in environmental degradation so severe that even when relocation and reorganisation policies were relaxed, it continued to affect indigenous populations. During this period 'all facets of aboriginal life [were] influenced by environmental degradation, [and] it is the traditional activities – reindeer herding, hunting, trapping, gathering and fishing which have been most affected'.[36]

Although Soviet environmental policy was, in effect, to have no policy at all aside from the exploitation of resources, it was cooperation on international environmental policy that provided the opportunity for political engagement with the other key states outside of the realm of military insecurities. This included participation in the *Antarctic Treaty 1959*, where Russia suspended territorial claims, but also the 1973 *Agreement on the Conservation of Polar Bears*. Incidentally, despite this late awakening to ecological consciousness and the myopic perspective on the exhaustibility of natural resources, it was the Soviet leader Gorbachev who initiated the advancement of international cooperation on the basis of environmental concerns in the Arctic.

An early step towards post-Cold War cooperation in the Arctic, Gorbachev suggested in his 1987 Murmansk speech to 'let the North Pole be a pole of peace'.[37] This speech lays out the elements that became the foundations of contemporary Russian Arctic policy: international cooperation, security, scientific research, the environment and Northern development. Yet even during the final days of the Cold War, this turn to cooperation came with caution that 'promoters of environmental regimes for the Arctic should be aware of this Soviet desire to limit the players in the Arctic to those with direct interests and activities there, keeping control in the hand of the Arctic eight'.[38] However, the Soviet Union would soon disintegrate, resulting in a change in the structure of the international system. The speech at Murmansk signalled to the other Arctic states that the future of Russian Arctic policy was inclined towards international cooperation, addressing the issues arising within the region and providing the framework for isolating Arctic issues between the relevant states.

The visual representations produced on postage stamps during this period continued to emphasise the Russian performances of sovereignty that communicated its perceived legitimate position of authority over the territories of the Arctic. This included significant emphasis on the role of icebreakers and their function of maintaining the NSR in several prominent series in 1976, 1977, 1978 and 1981. A second performance is the emphasis on the role of scientific exploration with images in 1955, 1958, 1962, 1977, 1979 and 1988, as well as continued emphasis on historic exploration, with 1949, 1957 and 1991 images featuring Bering and a 1959 image of Popov. Russia also draws some attention to the significance of the polar bear in a 1987 series.

Contemporary Russian Arctic policy

Following the breakdown of the Soviet Union and the transition to the Russian Federation, Russian Arctic policy has adapted both to the change in the character of the domestic political structure of Russia and to the transition in the structure of the international system. The loss of global predominance that Russia experienced in the shift away from the bi-polarity of the Cold War caused a crisis in Russian identity. This had implications for the ways that Russia communicates its Arctic engagement, including a significant amount of posturing taking place in its performances of sovereignty in the Arctic. Although the spaces over which Russia claims sovereignty in the Arctic have not changed on their maps, the rules around how sovereignty over this territory is claimed have changed. So even though official Arctic policy adheres to international law, the communication of territorial authority through performances of the old rules gives the impression that Russia is an aggressive and expanding Arctic state.

The key to understanding Russian Arctic policy is in the identification of national interests, delivered in the 2008 *Foundations of State Policy of Russian Federation in the Arctic for the period to 2020 and Beyond*. This policy document outlines Russia's national interests, issues and objectives in terms of its use as a resource base, focusing heavily on the development of the NSR, including 'the creation and development of the infrastructure and communications management system of the NSR to solve the problems of Eurasian transit'.[39] It identifies a desire to conserve the Arctic ecosystems and to maintain the Arctic as a zone of peaceful cooperation, while building a military 'capable of ensuring military security under various military-political scenarios'.[40] In addition to maintaining the integrity of Russia's historic Arctic claims, these policy elements give rationality to Russian state practice in the Arctic. This includes their involvement in international agreements such as the *Arctic Search and Rescue Treaty 2011*, the 2014 *Polar Code* and regional fishing moratoriums, but also in their Arctic military preparations, a source of geopolitical concern.

Fulfilling Gorbachev's invitation to launch international cooperation in the Arctic and transform it into a zone of peace, Russia has been an active participant in the creation of Arctic governance. This began with the 1991 *Arctic Environmental Protection Strategy* and the subsequent establishment of the Arctic Council. Russia also participated in the *Ilulissat Declaration 2008* where the Arctic states committed to abide by principles of international law, while finding solutions to regional issues outside of a formal Arctic Treaty. In 2010, Russia and Norway concluded a 40-year dispute over a maritime boundary in a mutually agreed bilateral treaty. This extensive cooperation together helps to address Arctic issues which Russia cannot manage alone, but also maintains control of the Arctic between the Arctic states as international interest in Arctic resources accelerates.

A critical element of Russian policy is its compliance with *UNCLOS 1982*, the framework which added new legal areas of maritime territory, lending the procedures to establish sovereignty over these spaces. Having ratified the convention in 1997, Russia has made the submission of their claims to the Commission on the Continental Shelf within the required 10-year deadline. Russia's initial submission in 2001 contained maps depicting the maritime borders as including the entire length of the median line between the US and Russia, and nearly all territory up to the North Pole on their western boundary.[41] This submission was returned for additional scientific evidence required by UNCLOS, but the revised 2015 submission still includes territory around the North Pole. Along with claims to the pole under UNCLOS, the Russian posturing in the flag planting exercise on the seabed of the North Pole in 2007 promotes its intention to maintain current boundary delimitations. While this was not a state-sponsored expedition, it was immediately celebrated in a double issue postage stamp *Arctic Deep-Water Expedition* released the same year.

Since the Age of Discovery, Russia has considered the Arctic area north of its *terra firma* as its territory. In each phase of the development of the international system it has implemented some form of control over the maritime region, from the banning of hunting when foreign users avoided paying Russian customs and taxes, through to the establishment of the NSR Administration that determines who is allowed passage through the maritime route. Considering this long-term control over the maritime Arctic, it is expected that Russia will maintain the desire to control the economic benefits of the region despite the changes to rules of territorial sovereignty. However, the adoption of the NSR as a major shipping lane from China to Europe is likely to encourage economic development for Russia and is thus in the national interest. The increase of traffic in this sensitive region also increases the possibility of industrial accidents and environmental damage.

The condition of the Arctic environment has been given a small platform in Russian policy and is being addressed through several different strategies, although there is much to be improved concerning internal environmental impact assessment practices. The first strategy is participation in environmental mechanism in the international sphere. In addition to environmental strategies associated with the Arctic Council, Russia is cooperating with the US Environmental Protection Agency and other states on the Arctic Military Environment Program. The aim of this programme is to assist Russia in the decommissioning of Cold War era nuclear arms and to clean up hazardous wastes previously dumped into the Arctic Ocean.[42]

Cooperation by Russia in international agreements, and specifically *UNCLOS 1982*, are merely a method to attract international investment in hydrocarbon industries, notwithstanding the impact of international sanctions following the annexation of Crimea. Russia's adherence to this legal regime and the legitimisation of its sovereignty gives foreign investors a sense of legal security in the designation of property rights for the territory in which these resources are located. However, the 2008 Russian Arctic policy document clearly indicates that the Russian Federation considers its boundaries to have been determined in the 1926 Soviet declaration on the sector principle. If these claims to the Commission on the Continental Shelf are denied, it is unlikely that Russia will change the territorial borders on maps.

Russia has drawn its international boundaries up to the North Pole in the pie-shaped wedge of the sector principle since 1926. Russia also considers its airspace to fall within this territory, making protest when this airspace is violated. Given the layers of territory which Russia has long embraced as part of its cultural and political heritage, it is unlikely that it will be willing to redraw the lines on its maps any time soon. These lines consistently appeared on military maps and propaganda posters throughout the Cold War period, often with a Soviet flag displayed at the North Pole. In this contemporary period, these performances of sovereignty emphasising this policy feature in the visual representations on postage stamps. These include images that continue to confirm the role of icebreakers in maintaining the NSR, such as the 2008 stamp celebrating the International Polar Year, while displaying the route used by icebreakers on a map of the Russian Arctic. It also continues in a 2009 series on icebreakers and a 2010 image celebrating the contributions of the polar scientist Yevgeny Fyodorov.

Conclusions

The visual representation of Russian policy in its postage stamps gives an interesting overview of the narrative of the Russian relationship with the region, reflecting the dynamics of its performances of Arctic sovereignty. The expansion of Russia over the Arctic, although at times failing in efficiency, is undoubtedly a significant source of wealth and prestige for this state. Throughout the duration of its sovereignty over the Arctic, Russia has had several constant policy objectives. The images found in these postage stamps emphasise Russian projection sovereignty and effective occupation over the maritime Arctic through historic exploration, the use of ice breakers and aeroplanes and, finally, through the conducting of scientific exploration.

In this long-term overview, Russian Arctic policy reveals itself to be complex, influenced both by trends in the structure of the international system and by transitions within its internal political situation. Often, its policy objectives and national interest reflect the structure of the international system, but at other times, the exigencies of the internal Russian political scene take precedence as they communicate perceived threats from other nations. This is seen especially in the increased use of scientific exploration as a political technology to confirm their presence in the maritime Arctic, when the policies of colonisation used elsewhere were impossible. Russian engagement with the Arctic throughout the three different political structures has maintained an impressive continuity as the Russian state has pursued politics of resource exploitation and effective occupation. Despite the focus on economic incentives throughout the circumpolar Arctic, Russia has been instrumental in developing the current international structures, found predominantly in the Arctic Council, critical to maintaining the Arctic as a zone of peace.

Notes

1 James Oliver, *The Bering Strait Crossing: A 21st Century Frontier Between East and West* (Exmouth: Company of Writers, 2006), p. 165.
2 Terence Armstrong, 'Russian Settlement in the North', *Arctic* 37, no. 4 (1984): 429–40, p. 432.
3 Yuri Slezkine, *Arctic Mirrors: Russia and the Small Peoples of the North* (Ithaca, NY: Cornell University Press, 1994), p. 13.
4 Raymond H. Fisher, *The Russian Fur Trade, 1550–1700* (Berkeley: University of California Press, 1943), p. 48.
5 John Tillotson, *Adventures in the Ice: A Comprehensive Summary of Arctic Exploration, Discovery and Adventure* (London: James Hogg & Son, 1869).
6 Oliver, *The Bering Strait Crossing: A 21st Century Frontier Between East and West*, p. 42.
7 Mary Foster and Steve Henrikson, 'Symbols of Russian America: Imperial Crests & Possession Plates in North America', *Alaska State Museum Concepts*, 2009, p. 2.
8 Theordore Lyman, *The Diplomacy of the United States* (Boston: Wells & Lilly, 1828), p. 287.
9 Ilya Vinkovetsky, 'The Russian-American Company as a Colonial Contractor for the Russian Empire', in *Imperial Rule*, eds. Alexei Miller and Alfred J. Rieber (Budapest: Central European University Press, 2004), p. 164.
10 Lydia T. Black, *Russians in Alaska 1732–1867* (Fairbanks, Alaska: University of Alaska Press, 2004), p. xiii.
11 The Parliament of the United Kingdom, 'The Boundary Between the British and Russian Empires on the North-West Coast of America' (The Diplomatic Review, March 23, 1859), p. 33.
12 Shelagh Grant, *Polar Imperative: A History of Arctic Sovereignty in North America* (Vancouver, Canada: Douglas & McIntyre, 2010), p. 91.
13 Piers Horensma, *The Soviet Arctic* (Abingdon, UK: Routledge, 1991), p. 12.
14 V. Yu. Alexandrov and V. Ye. Borodachev, 'History of the Northern Sea Route', in *Remote Sensing of Sea Ice in the Northern Sea Route: Studies and Applications*, ed. O.M. Johannessen (New York: Springer, 2007), p. 12.
15 Horensma, *The Soviet Arctic*, p. 15.

16 Horensma, p. 15.

17 Horensma, p. 25.

18 Timothy Taracouzio A., *Soviets in the Arctic* (New York: Macmillan Company, 1938), p. 320.

19 Matthew E. Lenoe, *Closer to the Masses: Stalinist Culture, Social Revolution, and Soviet Newspapers* (Cambridge, MA: Harvard University Press, 2004), p. 222.

20 John McCannon, *Red Arctic: Polar Exploration and the Myth of the North in the Soviet Union 1932–1939* (New York: Oxford University Press, 1998), p. 118.

21 John McCannon, 'Tabula Rasa in the North: The Soviet Arctic and Mythic Landscapes in Stalinist Popular Culture', in *The Landscape of Stalinism: The Art and Ideology of Soviet Space*, eds. Evgeny Dobrenko and Eric Naiman (Seattle: University of Washington Press, 2003), p. 243.

22 Frank J. Miller, *Folklore for Stalin: Russian Folklore and Psuedofolklore of the Stalin Era* (London: M.E. Sharpe, Inc, 1990), p. 45.

23 Terence Armstrong, *The Northern Sea Route: Soviet Exploitation of the North East Passage* (Cambridge: Cambridge University Press, 1952).

24 Armstrong, *passim*.

25 Harry Smolka P., *40,000 Against the Arctic: Russia's Polar Empire* (New York: W. Morrow & Co., 1937), p. 85.

26 James Forsyth, *A History of the Peoples of Siberia: Russia's North Asian Colongy 1581–1990* (Cambridge: Cambridge University Press, 1992), p. 316.

27 Cathleen Rineer-Garber, 'The Transition to a Free Market Economy in the Russian Far East: The Environmental and Social Consequences', in *Handbook of Global Environmental Policy and Administration*, eds. Denis Soden and Brent Steel (New York: Marcel Dekker, Inc, 1999), p. 396.

28 Lynne Viola, *The Unknown Gulag: The Lost World of Stalin's Special Settlements* (Oxford: Oxford University Press, 2007).

29 Clifford J. Webster, 'The Soviet Expedition to the Central Arctic, 1954', *Arctic* 7, no. 2 (1954): 58–80, p. 62.

30 Willy Østreng, 'Political-Military Relations Among the Ice States: The Conceptual Basis of State Behavior', in *Arctic Alternatives: Civility of Militarism in the Circumpolar North*, ed. Franklyn Griffiths (Toronto: Science for Peace, 1992).

31 Michael Marsden, 'Arctic Contrasts: Canada and Russia in the Far North', *International Journal* 14, no. 1 (1958): 33–41, p. 33.

32 Steven E. Millar, 'The Arctic as a Maritime Theatre', in *Arctic Alternatives: Civility of Militarism in the Circumpolar North*, ed. Franklyn Griffiths (Toronto: Science for Peace, 1992), p. 212.

33 Terence Armstrong, 'Northern Affairs in the Soviet Union', *International Journal* 19, no. 1 (1963): 40–9.

34 W.W. Nassichuk, 'Forty Years of Northern Non-Renewable Natural Resource Development', *Arctic* 40, no. 4 (1987): 274–84, p. 274.

35 Marian Leighton, 'Soviet Strategy Towards Northern Europe and Japan', in *Soviet Foreign Policy in a Changing World*, eds. Robbin F. Laird and Eric P. Hoffmann (New York: Aldine Publishing Company, 1986), p. 292.

36 Gail A. Fondahl, 'Environmental Degradation and Indigenous Land Claims in Russia's North', in *Contested Arctic: Indigenous Peoples, Industrial States, and the Circumpolar Environment*, eds. E.A. Smith and J. McCarter (Washington DC: University of Washington Press, 1997), p. 68.

37 Mikhail Gorbachev, 'The Speech in Murmansk at the Ceremonial Meeting of the Order of Lenin and the Gold Star Medal to the City of Murmansk' (Novosti Press Agency Publication House, 1 October 1987).
38 Gail Osherenko, 'Environmental Cooperation in the Arctic: Will the Soviets Participate?', *Current Research on Peace and Violence* 12, no. 3 (1989): 144–57, p. 150.
39 Russian Federation, 'Russia Federation's Policy for the Arctic to 2020' (Russian Federation, 2008), www.arctis-search.com/Russian+Federation+Policy+for+the+Arctic+to+2020#I._General_Provisions.
40 Russian Federation.
41 Russian Federation, 'Outer Limits of the Continental Shelf Beyond 200 Nautical Miles from the Baselines: Submission by the Russian Federation' (United Nations, 2001), www.un.org/depts/los/clcs_new/submissions_files/submission_rus.htm.
42 United States Environmental Protection Agency, 'Past Programs in Russia' (United States Government, 2013), https://archive.epa.gov/international/regions/web/html/russiapast.html.

5 Comparing performances of Arctic sovereignty

Introduction

Over the long twentieth century, the engagement with the Arctic by the US, Canada and Russia has transitioned and evolved. Throughout this time, they have found new ways to communicate and exercise effective occupation in their performances of sovereignty as they attempt to legitimise their ownership of the Arctic. For each of these states, there are similarities in policy objectives, in that they are all concerned with maintaining the legitimacy of their claims to this frontier of territorial development, but how they approach this goal differs, somewhat influenced by the extent of the spatial area over which they claim sovereignty. The differences appear to lie in whether they claim sovereignty only over land and statutory maritime spaces, where it is relatively standard practice to demonstrate sovereignty and rights according to the principles of international law, or whether these claims are over wider areas of maritime space, such as the North Pole. This chapter will discuss how these relationships emerge, and considers the quantitative relationship in the production of visual narratives in postage stamps. Through this quantitative analysis, the underlying policy trends emerge, revealing the prioritisation of policy agendas of importance to national priorities as projected in the Arctic postage stamps of these states.

Although for each of these states, the policies 'are premised on the four pillars of sovereignty, economic development, environmental protection and security, the latter three, in fact, are supportive of the overarching aim of establishing or maintaining sovereignty'.[1] Each of these states manifest these interests in different ways, in part because of the internal motivations and identities of the states given the specific conditions of their relationship to the Arctic, but also sometimes influenced by domestic factors internal to these states, such as personality politics. Alongside the dynamics of national interests and internal conditions, these policy areas are projected onto the two different levels of politics over national and international spaces.

The result of considering the entirety of Arctic policy and engagement between these three states across the long twentieth century is that it is possible to view the evolution and transitions in the policy approaches of the US, Canada and Russia in response to the changing structure of the international system. Some of the broader global pressures of this century include changes in the solidification of the state system emerging from the fragmentation of imperialism and resulting decolonisation, the implications of security agendas, first in the world wars and then followed by the power bi-polarity of the Cold War, the significant expansion of the legal categories of maritime territory in the law of the sea, and the rapid transformation of technological capabilities. While none of these changes specifically targeted the Arctic region, they certainly impacted the policies of states in the ways that they engaged with the Arctic as they responded to the new and developing norms.

Within each of these states, there are also differences in the way that they have formed their relationship with the Arctic due to internal policy objectives relative to the broader national interest of the state. At times, the state-based approaches to the Arctic have been heavily contingent upon the agendas important to specific moments of leadership. However, what is specifically curious is that despite these variations in approaches between the domestic and the international, and the usually very clear divisions between foreign and domestic policy within states, these Arctic policies do not follow these usual lines of demarcation. Instead, the policies and strategy documents of the US, Canada and Russia towards the Arctic are heavily blurred between internal and foreign policy, producing broad approaches for engagement with the region that straddle both domestic and international spaces, areas over which they make claims to establishing or maintaining sovereignty. This is not to be considered as evidence of imperial land-grabbing. It is rather to emphasise that the exceptional space of the Arctic is not only due to its sacredness in historical and cosmological lore, but also due to its legal ambiguities as an ice-covered region, where ice is sometimes treated as land, rather than as sea.

Comments on methodology

As the silent messengers of the state, postage stamps deliver messages charged with politicized narratives, as they advertise and provide publicity. These images are the 'ways in which the state visually presents (or misrepresents) its history, culture, society and their place in the world'.[2] As the US, Canada and Russia present the different performances of sovereignty they have used in their policies towards the Arctic, these images have subtly encouraged the legitimisation of their sovereignty over the Arctic to both domestic and international audiences. These results are 'achieved through the dissemination of images that suggest a familiarity of geographic spaces, repetitive use of indigenous iconography, and pictorial representations of effective state administration over the region'.[3] The repetition of these themes 'further consolidates the process of delivering the message', making the messages familiar and comfortable.[4] This technology has been used by the US, Canada and Russia to communicate policies that correspond to the exigencies of their Arctic sovereignty relevant throughout these periods.

Within the academic discourse on postage stamps there is consensus that postage stamps are representatives of the government and that they contain messages, some straightforward, such as the country of issue and the amount paid for postage. However, some messages are more versatile and are subliminally influenced by the use of national symbols or other graphic references to national history, culture and geography. Stamps 'reflect ideologies, aspirations and values, attesting to political, social and cultural ideas and aesthetic tastes. When closely inspected, stamps often prove to convey much more than meets the eye.'[5] From these images 'meaning [can be] derived from the nature of the whole image and circumstances of its creation'.[6] Postage stamps are useful as propaganda as they reinforce national identity to a domestic audience, but stamps are also distributed to an international audience who are informed of the characteristics of the issuing country.

This study covers the entirety of the postage stamps issued by the US since 1847, by Russia (and the Soviet Union) since 1857, and by Canada since 1851 until 2010, using postal archives, museums and catalogues. From a study of these postage stamp collections, a total of 212 postage stamps representing the Arctic were identified: 36 from the US, 77 from Canada and 99 from Russia.

Methodology for this research and analysis of each stamp has been informed by the approaches used by Hammet, Childs, Raento and Brunn, Hoyo, and Frewer in their studies of South African, South American, Finish, Soviet, Mexican and Japanese stamps.[7] By considering the images of Arctic stamps, including who or what categories of items were depicted, five key iconographic themes were identified by organising images of stamps into relevant categories. These themes were then 'isolated to identify ways in which spaces, places and people were constructed and depicted, overtly and covertly, through expressions of power and ideology'.[8] This process was used to fulfil the query of the research questions asking what has been depicted on the Arctic stamps of the US, Canada and

Russia. A discussion of the way that this production occurs across the eras and the differences in representations between states follows.

Producing visual representations of policy

Throughout the four periods that this research has discussed, from the Age of Discovery through to the contemporary, post-Cold War period, there are differences in the production of the visual representations in the postage stamps of the US, Canada and Russia. Although it is relatively easy to see that the quantities have changed for all the states across all the periods, the question to consider is what accounts for the different volumes of production in the different eras.

During the Age of Discovery, postage stamps only emerged as an administrative technology near the middle of the nineteenth century. Most early postage stamps merely depict the heads of state, which is one of the features that lends legitimacy and authority to this technology. As a result, it took some time for the trends of postage stamp production to migrate away from featuring the head of state to instead making commemorative representations of interest to the nation. This trend accounts for the slow production of Arctic images in this period. Given this, stamps of Canada

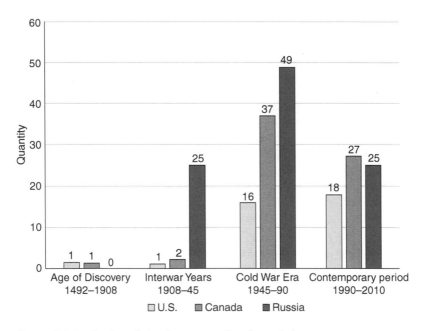

Figure 5.1 Production of visual representations by period.

are unique in that their first postage stamp featured a beaver rather than the reigning British monarch. The first US stamps had two joint issues, one featuring the first president, George Washington, and the other featuring the first post-master general. All Russian stamps in this period featured the symbols of the empire, including the Tsar and the double headed eagle.

In the Interwar Years there is a much more notable variation in the quantity of production of postage stamps across the states. The continued low population of US stamps can be explained by the very stable condition of sovereignty over Alaska. Although Alaska was merely incorporated as a territory, the nature of the acquisition by treaty and purchase gave little opportunity to doubt ownership, even if the effective occupation of the territory was sometimes ineffectual. Canadian stamps still largely reflected the country's status as part of the British Commonwealth, and although the legitimacy of its sovereignty over the Arctic was in question, it had yet to begin its campaign to communicate intentions towards the region, beyond its position on the sector principle. The significant acceleration by Russia is understood in the context of the emergence of the Soviet Union after the Russian Revolution. This is the period when the new government issued stamps that reflected the new political situation and ideology of the Communist takeover in the broader contexts of nation-building when 'the new postal authorities were seeking strategies to enhance Russian nationalism', as well as the significance of the Arctic to the Stalin administration.[9]

During the Cold War Era, the production of postage stamps across the states increased significantly for both the US and for Canada, but only just doubled for Russia. This represents the changing trends in the use of postage stamps around the world as a method of commemorating 'the events, peoples and places that impinge on a nation's consciousness'.[10] A significant trend in Russia during the period was the communication of 'mastery over nature', where both outer space and the Arctic were convenient other worlds for displaying Russian prowess. For Canada, this was the era when they became most conscious of the crisis of legitimacy over their Arctic sovereignty, in part due to American military presence, and under the Trudeau administration they began to communicate at large their intentions of maintaining sovereignty in the region. For the US, there is also an increase in communicating their Arctic sovereignty, in a small part brought about by the brief occupation of Alaskan islands during the Second World War, but also due to the shift in US strategy to communicate their apparent turn to using environmental protection as a strategy for projecting legitimacy.

In the Contemporary Period, the quantity of production across these states has mostly levelled out, although Canada has surpassed Russia. This can mostly be explained by the significant amount of emphasis by Canada on Arctic sovereignty under the Harper administration.

Themes and trends in the visual representations of policy

When the Arctic postage stamps of these states were categorised according to the images on the issues, five clear themes emerged: nature, indigenous people, effective occupation, exploration and finally, maps, flags and territory. These themes were selected to align with the policy trends in Arctic postage stamps to allow 'for critical analysis of the social and power relations framing philatelic iconography'.[11] For each of these three states, the theme with the most emphasis varies, greatly affected by the internal policy preferences of the state.

In this graph, the emphasis on nature emerges as a key priority of the US, a trend which strongly correlates with their tendency to project themselves as good stewards of the environment. This theme dominates the visual representations of the US, while the three categories of indigenous people, exploration and maps, and flags and territory are all produced in equal quantities. The least dominant theme in US production is effective occupation, which is expected given the confidence of the US in their ownership of Alaska.

For the visual representations produced by Canada, the theme of indigenous people is the most communicated policy preference in their postage stamps. The strength of this preference is understood given the long-term agenda of Canada emphasising the enfranchisement of the Inuit as citizens of the nation. It is interesting that this projection might have a second intent in communicating this enfranchisement back to the Inuit, given a survey finding that indigenous communities have the highest positive acceptance of national symbols amongst Canadians.[12] Following on from this, its policy preferences closely follow the trends of US postage stamps, with nearly equal production in the themes of nature, exploration and maps, and flags and territory, while effective occupation is the lowest ranked theme. Given the emphasis on Arctic sovereignty in government discourse in the last period, this low emphasis on effective occupation is explained by the large shift in Canadian policy from effective occupation through development to a focus on indigenous enfranchisement after the Interwar period.

For Russia, effective occupation is the theme with the greatest consideration, with nearly all the effective occupation images focusing on maritime administration. Some of these images are celebrating scientific achievements at the North Pole, but most of these images are of ice breakers. This importance aligns with the fact that Russia has the world's largest fleet of ice breakers and these vessels are essential for navigating the NSR. The next most significant theme for Russia is exploration. This was predominantly represented in images of historic ships and celebrating the explorers of the Russian Arctic. The least significant theme for Russia is its indigenous people. This strongly correlates with the low priority of indigenous rights and issues within the country.

When these themes are considered with a comparative lens against the other states, it becomes clear how these states' policy preferences are communicated in their visual representations. Building on the information on themes in Figure 5.2, Figure 5.3 continues to look at the themes as produced within each state, but in addition adds quantitative analysis by considering the mean and standard deviation of the production of these images. This reveals a clear hierarchy within the themes, as it stacks the production of themes, showing comparative state by state preferences.

Looking first at the theme of nature, the US is a clear outlier in this theme, with both Russia and Canada having nearly the same production weighting. This shows the relative significance of the message for US policy preferences against the Canadian and Russian trends. In some respects, this low ranking is surprising in Canadian production given their internal position on legitimising their control over the Northwest Passage through environmental protection; however, the lower ranking of nature compared with indigenous people is understood given the significant priority that the Inuit have taken in the Canadian agenda over the last two

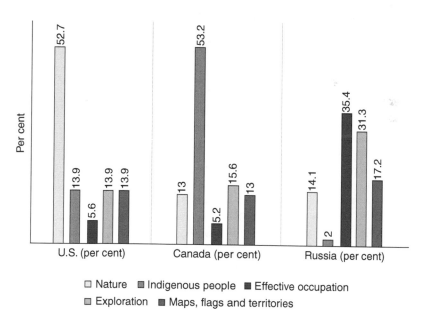

Figure 5.2 Visual representation of policy by state.

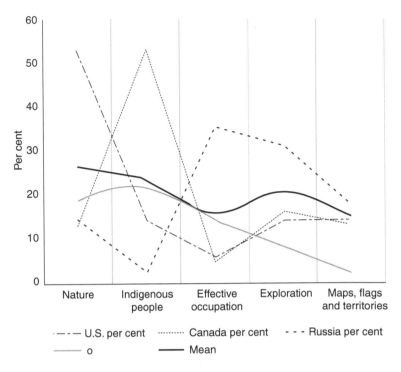

Figure 5.3 Comparative policy preferences.

periods. The relative low ranking of nature for Russia is understood against the backdrop of weak environmental protection practices – especially in the context of oil and gas pipeline development crossing the Arctic region when these projects are usually approved in advance of environmental impact assessments. However, given the importance of the polar bear in Russian iconography, this has helped to put Canadian and Russian nature productions on equal levels.[13]

In the theme of indigenous people, Canada holds the production extreme, while the theme is middle ranking for the US and almost non-existent within Russian production of postage stamps. These positions show very clearly the relative importance of indigenous issues and the role of indigenous people in the project of Arctic sovereignty in these states. For Canada, the Inuit are extremely significant to establishing and maintaining Canadian Arctic sovereignty in the enfranchisement of the Inuit who have lived in the Arctic since 'time immemorial'. While for the US, indigenous peoples play almost no role in the maintenance of Alaska, reflecting the guardianship status of native Americans throughout US history. For Russia, the very low production reflects the legal discrimination towards indigenous peoples inherent in the Russian political system.[14]

Under the theme of effective occupation, Russia is the clear outlier for production, while for Canada and the US, this theme rates lowly. This occupation has a significant presence for Russia given their continued emphasis on demonstrating their sovereignty over the NSR, a maritime passage that stands to become a significant feature of global shipping trade as the Arctic climate warms. The relatively low frequency of production for the US is explained in that they have few sovereignty insecurities over their Arctic territory. For Canada this is explained by their reliance on other policy methods for communicating their claims to Arctic sovereignty.

The theme of exploration also features strongly in Russian policy emphasis, a trend which is relatively consistent throughout all the periods. This is in contrast to the lower Canadian production, where exploration of the Arctic is a historical method used to emphasise the role of Great Britain in exploring the region. Although this may have established initial rights of the Arctic for Canada, they acquired this through imperial transfer. For the US, the policy of exploration of the Arctic primarily has significance in celebrating the achievement of Americans in expanding polar knowledge, contributing more to national security than to claims to sovereignty in the region.

The category of maps, flags and territory is the point at which these states nearly merge in their production. In this category, nearly all the Russian stamps feature a flag at the North Pole, where for Canada, there were images of flags over maritime spaces and natural resources. The US images in this category largely feature incorporation of territory into national administrative mechanisms.

The quantitative analysis emphasises interesting trends regarding the production of postage stamps by these states across the themes. On the theme of nature, the US is significantly higher than the mean, where Russia and Canada are slightly low and the standard deviation shows that the US performance in this area is outside of normal conditions. On the theme of indigenous people, Canada is significantly higher than the mean, while the US is slightly lower and Russia is significantly lower. The standard deviation for nature shows there are significant margins for this category, with both Canada and Russia lying far outside the norm. For both effective occupation and exploration Russia places higher than the mean, with the US and Canada slightly lower. The standard deviation for these means shows that the performances of states with regards to this category is slightly outside of standard margins, but that these positions are not extreme. The only point at which these states are near the mean is in the category of maps, flags and territory, and this is the category with the lowest standard deviation, showing a clear and stable trend between these states on this theme. This is the point at which there is the most coherence and agreement in the performance of policy across these states.

Variations across state administrative periods

This set of graphs portrays how production of postage stamps changed across the different leadership administrations of the US, Canada and Russia. The graphs show three different measurements for each administration period: the range of years, the average population of postage stamps and the per cent total production of postage stamps. The most important data of this graph is the resulting average population calculations, which shows how postage stamp production differed throughout the different administrations, relative to each state. This information allows the consideration of the weight of postage stamp production beyond the mere units of production. This data shows a distinct period in which the communication of Arctic sovereignty was most significant across all states, this is the time period that aligns with the discovery of oil reserves in the Arctic maritime. The second highest population rating for all states also

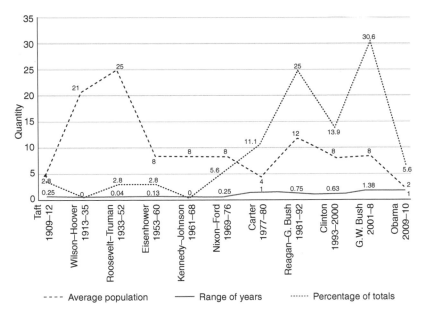

Figure 5.4 Production of visual representations throughout US administrations.

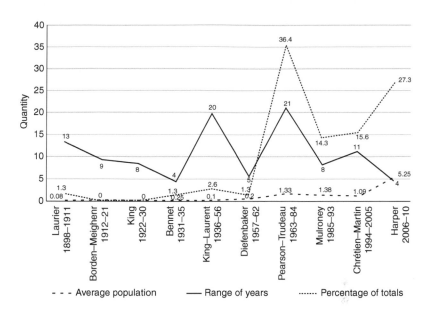

Figure 5.5 Production of visual representations throughout Canadian administrations.

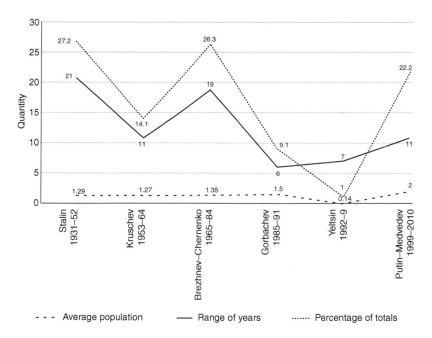

Figure 5.6 Production of visual representations throughout Russian administrations.

overlaps with the period around the end of the Cold War, and the time at which *UNCLOS 1982* was ratified, when there were two significant rule changes for the Arctic region.

Conclusions

This quantitative consideration of the postage stamps of the US, Canada and Russia has shown how the production of these images relates to the grander narratives of Arctic policy and the performances of sovereignty throughout these states. The results of this quantitative analysis demonstrate that within each of these states there are underlying domestic policy trends, sometimes aligning with the interests of specific administrations, but also aligning with transitions within the rules of the international system and physical understandings of the Arctic region. For the US, this correlates with their projection as environmental stewards of natural resources through images of national parks, flora and fauna – even if only within Arctic international politics. The Canadian priority exists in promoting the indigenous connection of their Arctic sovereignty through emphasizing the importance of the Inuit into the historical narrative of Canadian state-building, a strategy maximised under the Harper administration. Meanwhile, Russian priorities were emphasised in projections of Russian skill of mastery over nature in their history of polar scientific expeditions and Arctic operational prowess, beginning in Stalinist projects and continuing through to the Putin administration.

Notes

1 Corine Wood-Donnelly, 'Messages on Arctic Policy: Effective Occupation in the Postage Stamps of the United States, Canada and Russia', *Geographical Review* 107, no. 1 (2017): 236–57, https://doi.org/10.1111%2Fj.1931-0846.2016.12198.x, p. 8.

2 Pauliina Raento and Stanley D. Brunn, 'Visualizing Finland: Postage Stamps as Political Messengers', *Geografiska Annaler Series B, Human Geography* 87, no. 2 (2005): 145–63, p. 145.

3 Wood-Donnelly, 'Messages on Arctic Policy: Effective Occupation in the Postage Stamps of the United States, Canada and Russia', p. 5.

4 Jack Child, *Miniature Messages: The Semiotics and Politics of Latin American Postage Stamps* (Durham, NC: Duke University Press, 2008), p. 4.

5 Ami Ayalon, 'The Hashemites, T.E. Lawrence and the Postage Stamps of the Hijaz', in *The Hashemites in the Modern Arab World: Essays in Honour of the Late Professor Uriel Dann*, eds. Asher Susser and Aryeh Shmuelevitz (Abingdon, UK: Frank Cass & Co, Ltd, 1995), p. 16.

6 Douglas Frewer, 'Japanese Postage Stamps as Social Agents: Some Anthropological Perspectives', *Japan Forum* 14, no. 1 (2002): 1–19, p. 7.

7 See: Daniel Hammet, 'Envisaging the Nation: The Philatelic Iconography of Transforming South African National Narratives', *Geopolitics* 17, no. 3 (2012): 526–52. Raento and Brunn, 'Visualizing Finland: Postage Stamps as Political Messengers'. Henio Hoyo, 'Fresh Views on the Old Past: The Postage Stamps of the Mexican Bicentennial', *Studies in Ethnicity and Nationalism* 12, no. 1 (2012): 19–44. Frewer, 'Japanese Postage Stamps as Social Agents: Some Anthropological Perspectives'.

8 Hammet, 'Envisaging the Nation: The Philatelic Iconography of Transforming South African National Narratives', p. 528.

9 Stanley D. Brunn, 'Stamps as Messengers of Political Transition', *Geographical Review* 101, no. 1 (2011): 19–36, p. 29.

10 Igor Cusack, 'Tiny Transmitters of Nationalist and Colonial Ideology: The Postage Stamps of Portugal and Its Empire', *Nations and Nationalism* 11, no. 4 (2005): 591–612, p. 593.

11 Hammet, 'Envisaging the Nation: The Philatelic Iconography of Transforming South African National Narratives', p. 528.

12 Government of Canada, 'Report on Canadian Identity' (Ottawa: Statistics Canada, 2015), www150.statcan.gc.ca/n1/en/pub/89-652-x/89-652-x2015005-eng.pdf?st=TiZCzUsc.

13 Oleg Riabov and Tatiana Riabova, 'The Remasculinization of Russia?', *Problems of Post-Communism* 61, no. 2 (2014): 23–35.

14 Alexandra Xanthaki, 'Indigenous Rights in the Russian Federation: The Rights Case of Numerically Small Peoples of the Russian North, Siberia, and Far East', *Human Rights Quarterly* 26, no. 1 (2004): 74–105.

Conclusion

Performances of sovereignty are instrumental in shaping understandings and acceptance of the legitimacy of states to govern over specific geographical spaces. These performances are not always evidence for actual and absolute sovereignty. However, they help to create the essence of authority and to assist states in negotiating the transition between *terra communis*, territory over which no sovereignty exists, to territory that is framed within the bordered power container of the state. As such, performances of sovereignty facilitate the state's communication of policy, words and practices that together combine to become the process that secures the coveted territorial authority.

It is quite astounding to think about how the earth's land, once owned by all men in common, has become an instrument of power and materialised as private property, subdivided into the authoritative domains of a limited number of states. This transformation has, of course, developed over the temporal span of centuries and with this evolution, the concept of sovereignty has been critical in defining this metamorphosis that coincides with the development of the state system. As a concept of authority, sovereignty has its origins in the relationship between the ruler and the ruled, but over time this has expanded to describe also the relationship of a state with a given territorial space.

The landed territories of the circumpolar Arctic were subjected to this evolution of sovereignty at a time when the authority of the state and the modern international system was rapidly expanding across the globe. As a result, the Arctic was absorbed into the sovereign domain of a limited number of state actors who conducted performances of sovereignty in ways that communicated their compliance with the socio-legal rules of territorial conquest. This was first by discovering the territory and then by effectively occupying the land.

However, given the extreme conditions of the Arctic, both in its location at a distant point of the globe and the climatic conditions that accompany this geographic situation, the incorporation of the Arctic into the sovereign territories of a few states was hampered by the ice and cold. More importantly, it occurred late in the process of imperial state expansion and by the time the many Arctic lands had been discovered, the rules were changing. This forced states to modify the methods they used to ensure their authority over the Arctic was certain, as well as changing the methods used to communicate their authority over the region in a way that would be recognised by other states.

In arriving at the conclusions of this research, the concepts of performativity and speech acts are critical in considering the evolution of territorial authority in the Arctic. Because states and the bounded territories which they claim are socially constructed, they have been created through performances and through the repetition of practices that reinforce the rules of the international community of states. Bounded territory is merely the material manifestation of the legitimisation of this authority, and sovereignty over the territory of the Arctic is a manifestation of spatio-legal legitimisation of state authority in this specific geographic space.

The Arctic is a space that has been materially and discursively performed through words produced as acts of diplomacy and policy, and through the repeated practices of effective occupation. By performing the rules of sovereignty, the Arctic states have written the existence of their authority and legitimacy over this space into being. The story of this space could have been produced differently; it could have retained its status as *res communis*, territory of which mankind is sovereign, and preserved for the common heritage of the world. Instead, it has been written into the history of conquest and territoriality, a process of the legitimate exploitation of people, resources and space.

A core purpose of this book has been to narrate the record of this process by explaining about the rules of territorial sovereignty that have come to place the geography of the Arctic within the ownership of specific states. Complementary to this explanation is the narrative of how the US, Canada and Russia responded to changes in the rules of territorial sovereignty as they engaged with the Arctic. This has included responding to changes such as the expansion of maritime sovereignty, the rise of environmental concerns and indigenous rights as well as the ways in which technology impacted how man interfaces with the Arctic. All of these changes having occurred in the timespan of a century has meant that these states have had to adapt their performances of sovereignty to the new circumstances as they evolved inchoate titles over the region into recognised and legitimate sovereignty.

The record of these performances and many of the policies that the US, Canada and Russia have used to legitimise their authority over the Arctic are illustrated in the postage stamps produced by these states. The study of postage stamps reveals trends that are consistent across the performances of these states, showing that they are using similar methods and practices to generate the legitimacy of their sovereignty. This demonstrates that this process of legitimisation is reciprocal and co-constituted. While these states have responded to the changing rules of the international system in ways that reflect the internal agendas and national interests of their domestic constitutions, there is also a great deal of consistency in the methods used specifically to frame the sovereignty of the Arctic as legitimate extensions of the littoral states.

To make evolution of territorial authority in the Arctic more complex, the sovereignty in the Arctic has two different geographic compositions, land and sea. This is not merely a physical divide, it is also a legal divide. While performances of sovereignty over the landed spaces of the Arctic has been relatively straightforward, it is the sovereignty over the maritime spaces that has made the ongoing performances by the Arctic states more difficult to follow, especially since the conditions of the Arctic make it possible to treat the frozen sea as land. It is this physical environment, combined with the mixed uses of this space, that has made states respond with extraordinary performances of sovereignty in the Arctic, including the use of scientific research bases on ice platforms and environmental protection over areas internally considered as *mare clausum*.

The crux of the rationale for the performances of sovereignty over the territories of the Arctic is in part the maintaining of the historical legacies of exploration and the glory that accompanied the discovery of this remote part of the earth. It is, however, the quest for resources that provides the ultimate motivation for establishing sovereignty over the Arctic. The incentives for Arctic development were found first in maritime routes, followed by the fur trade, whale oil, mineral resources and now again, shipping routes and hydrocarbon extraction. As the Arctic is always framed as a frontier of resource opportunity, the establishment of sovereignty has never been about extending the social contract to Arctic peoples, rather it has always been about achieving access to the economic incentives inherent within the region.

It may be that there is an ongoing scramble for the territories of the Arctic, but only if this is defined as a glacial process. The establishment of Arctic sovereignty is an ongoing and lengthy process that responds to normative changes in the international system. Flag planting in maritime spaces, cartographic mapping or even giving Santa a passport may not now be accepted as legitimate performances of sovereignty. They are, however, still useful techniques because they communicate the intent of the state to a global audience. It is especially important when considered together as part of a constellation of performances of sovereignty over time and space.

The concept of sovereignty is yet evolving and the legal spaces of territory are also likely to change in the future. A century ago, the extended maritime spaces that states can now claim as territory did not exist. It is currently only speculation which legal territories may be created in the future or what practices will constitute effective occupation of these territories, but environmental stewardship and performance of legal responsibilities are already emerging as recognised methodologies in the Arctic. It is therefore critical for the US, Canada and Russia to continue communicating their intent towards maintaining effective occupation of the Arctic so that as the rules evolve, they can demonstrate long-term continuity in their engagement with this geographic space, legitimising their authority through these performances of sovereignty.

Index

Page numbers in *italics* denote figures.